NUMBERS IN PRESENCE AND ABSENCE:
A STUDY OF HUSSERL'S PHILOSOPHY OF MATHEMATICS

PHAENOMENOLOGICA

COLLECTION FONDÉE PAR H.L. VAN BREDA ET PUBLIÉE
SOUS LE PATRONAGE DES CENTRES D'ARCHIVES - HUSSERL

90

J. PHILIP MILLER

NUMBERS IN PRESENCE AND ABSENCE:

A Study of Husserl's Philosophy of Mathematics

J. PHILIP MILLER

NUMBERS IN PRESENCE AND ABSENCE:

A Study of Husserl's Philosophy of Mathematics

1982

MARTINUS NIJHOFF PUBLISHERS
THE HAGUE/BOSTON/LONDON

Distributors:

for the United States and Canada
Kluwer, Boston, Inc.
190 Old Derby Street
Hingham, MA 02043
USA

for all other countries
Kluwer Academic Publishers Group
Distribution Center
P.O. Box 322
3300 AH Dordrecht
The Netherlands

Library of Congress Cataloging in Publication Data

Miller, J. Philip.
Numbers in presence and absence.

(Phaenomenologica ; 90)
Bibliography: p.
Includes index.
1. Husserl, Edmund, 1859-1938. 2. Mathematics--Philosophy. I. Title. II. Series.
QA8.4.H87M54 1982 512'.72 82-8044
ISBN 90-247-2709-X AACR2
ISBN 90-247-2709-X (this volume)
ISBN 90-247-2339-6 (series)

PRINTED IN THE NETHERLANDS

In memory of my father,
EDWARD D. MILLER

ACKNOWLEDGEMENTS

A substantial portion of the research on which this study is based was carried out at the Husserl Archives in Cologne, West Germany. I am grateful to the staff of the Archives, and in particular to Professor Elisabeth Ströker, the Director, for their assistance.

My stay in Cologne was made possible by a grant from the Fulbright Commission for Educational Exchange, which I hereby gratefully acknowledge.

I also wish to express my special thanks to those who read and commented upon earlier drafts of this work, especially John Brough of Georgetown University and Thomas Prufer of The Catholic University of America. My deepest thanks however go to the Reverend Robert Sokolowski, also of The Catholic University of America; his stimulating suggestions and patient criticisms provided guidance for me at every stage of my work on this project.

CONTENTS

CHAPTER I: THE EMERGENCE AND DEVELOPMENT OF HUSSERL'S 'PHILOSOPHY OF ARITHMETIC' 1

1. Historical Background: Weierstrass and the Arithmetization of Analysis 1
2. Husserl's First Stage: Analysis as a Science of Number 2
3. Husserl's Second Stage: Analysis as a Formal Technique 10
4. Husserl's Third Stage: Analysis as Manifold Theory 15
5. The Problem of Psychologism in Husserl's Early Writings 19

CHAPTER II: HUSSERL AND THE CONCEPT OF NUMBER 31

1. The Definition of Number 31
2. The Origin of Number as a Phenomenological Problem 34
3. The Origin of Number in Husserl's Eearly Writings 37

CHAPTER III: THE PRESENCE OF NUMBER 45

1. Sensuous Groups 45
2. Explication 51
3. Comparison 55

CHAPTER IV: NUMBERS AS IDENTITIES IN PRESENCE AND ABSENCE 65

1. Intending Numbers in Their Absence 65
2. The Unity of Number 69
3. The Unity of Large Numbers 76
4. Sedimented Number Meanings 79

CHAPTER V: THE SENSE OF ARITHMETIC 89

1. Ideal Numbers 89
2. The Formal Character of the Concept of Number 98
3. Arithmetic as Formal Ontology 100

CHAPTER VI: THE SENSE OF ANALYSIS 109

1. The Algebraization of Arithmetic 109
2. Theory Forms and Manifolds 113
3. Analysis as Manifold Theory 120
4. Husserl's Attempted Justification of Analysis 125

CONCLUSION 135

NOTE ON ABBREVIATIONS 139

BIBLIOGRAPHY 141

CHAPTER I

THE EMERGENCE AND DEVELOPMENT OF HUSSERL'S 'PHILOSOPHY OF ARITHMETIC'

1. HISTORICAL BACKGROUND: WEIERSTRASS AND THE ARITHMETIZATION OF ANALYSIS

In the preface to his early work *Philosophie der Arithmetik*, Husserl boldly announced his intention to make a contribution to that 'desideratum of centuries,' a 'true philosophy of the calculus' (PA 7).[1] Despite the title of the work, it was not arithmetic in the usual sense, but rather calculus or 'analysis' itself – that great achievement of modern mathematics, and the field in which Husserl himself had received a doctorate several years earlier[2] – which stood at the center of his philosophical concern. The problems addressed in the work had occupied Husserl for several years prior to the publication of PA, and they would continue to be of interest to him until the very end of his career. In this study we will examine Husserl's treatment of these problems and of other problems which developed out of them. Prior to beginning our study of Husserl's own ideas, however, it will prove helpful to glance briefly at the historical context in which they arose.

The area of mathematics known as analysis originated in the seventeenth century through the epoch-making discoveries of Newton and Leibniz. The calculating techniques they developed made it possible to deal successfully with a great many problems that had previously been insoluble. These techniques proved especially important, of course, in the development of the mathematical sciences of nature.[3] It seems, however, that very little attention was paid at first to the theoretical foundations of the calculus. As one survey of the development of this discipline comments:

> It was natural that this wide and amazing applicability of the new subject should attract mathematical researchers of the day, and that papers should be turned out in great profusion with seemingly little concern regarding the very unsatisfactory foundations of the subject. It was much more exciting to apply the marvelous new tool than to examine its logical soundness, for, after all, the processes employed justified themselves to the researchers in view of the fact that they worked.[4]

See notes at the end of this chapter.

To be sure, the theoretical obscurities associated with the calculus did not go entirely unnoticed. The concept of the 'infinitesimal,' or of an infinitely small quantity, seemed especially troublesome to some, even though it also appeared to be essential to the calculus. Berkeley referred derisively to Newton's 'fluxions' as the 'ghosts of departed quantities,' while Voltaire described the calculus as an art of 'numbering and measuring exactly a thing whose existence cannot be conceived.'[5] But through most of the eighteenth century, mathematicians themselves remained generally unconcerned by these obscurities. They seem rather to have taken to heart the advice which d'Alembert gave to a hesitant mathematical friend: 'Allez en avant, la foi vous viendra!'[6]

In the nineteenth century, however, a new attitude emerged. Mathematicians themselves grew concerned about the difficulties regarding the fundamental concepts of the calculus and began to seek a firmer foundation for this important discipline. One of the most prominent of these mathematicians was the brilliant and influential Berlin professor Karl Weierstrass. Although major theoretical steps had been taken earlier in the century, it was Weierstrass who finally developed the resolution which is accepted by mathematicians even today. What Weierstrass advocated has been described as 'a program wherein the real number system itself should first be rigorized; then all the basic concepts of analysis should be derived from this number system.'[7] It is this program which subsequently came to be known as the 'arithmetization of analysis.' Through arithmetization, it was felt, all of the obscurities associated with the calculus could be eliminated. If the differential, for example, could be derived from the properties of the real number system, there would no longer be a need for the mysterious notion of an infinitely small quantity. Weierstrass and his followers eventually showed that this could be done not only for the differential, but also for other concepts of analysis as well. And thus, in the words of one commentary, 'today it can be fairly said that classical analysis has been firmly established upon the real number system as a foundation.'[8]

The cornerstone of Weierstrass' program was of course the rigorous development of the real number system itself. Mathematicians had spoken of 'rational,' 'irrational' and 'negative' numbers for some time prior to Weierstrass' work. But generally speaking, the understanding of real numbers had been based upon the intuitive notion of a number line. Weierstrass, however, sought to dispense with all such 'geometrical intuitions' and build up the system of real numbers in a purely formal manner. Beginning with a rigorous development of the system of positive whole numbers, he proceeded to generate the system of integers in general, the system of rational numbers and, finally, the system of real numbers itself, through a series of precisely formulated steps.[9] In this way, it was thought, the number system which was fundamental to analysis as a whole could be made secure.

We must assume that the young Husserl was thoroughly familiar with Weierstrass' work on the foundations of analysis. He had studied mathematics in Berlin from 1878 until 1881. In each of the six semesters he spent there, he had attended Weierstrass' lectures and taken extensive notes.[10] And after earning his doctorate in Vienna in 1883, he had returned to Berlin for a brief period to serve as an assistant to Weierstrass.[11] The impact which this extended association with the great mathematician had upon Husserl's intellectual development can scarcely be exaggerated. Husserl himself later said that this study with Weierstrass had 'had a lasting influence' on him.[12] In 1929, on the occasion of his seventieth birthday celebration, Husserl is reported to have mentioned 'two of his former teachers: Franz Brentano, of course, and, with particular emphasis, Karl Weierstrass.'[13] And Husserl's wife recalled that 'Weierstrass had the most profound influence on him [Husserl]. He used to say sometimes, 'I got the ethos of my scientific striving from Weierstrass.''[14]

But Weierstrass was much more than a personal inspiration to the early Husserl. For it was he, it seems, who first kindled Husserl's interest in the philosophy of mathematics.

> It was my great teacher Weierstrass [Husserl later reported] who, through his lectures on the theory of functions, aroused in me during my years as a student an interest in a radical grounding of mathematics. I acquired an understanding for his attempts to transform analysis – which was to such a very great extent a mixture of rational thinking and irrational instinct and knack – into a rational theory. His goal was to set out its original roots, its elementary concepts and axioms, on the basis of which the whole system of analysis could be constructed and deduced by a fully rigorous, throughly evident method.[15]

Clearly, what Husserl is referring to in this passage is Weierstrass' attempt to 'arithmetize' analysis, to unify analysis and elementary arithmetic in a single, comprehensive *arithmetica universalis*.

Weierstrass' published writings are of course concerned primarily with the purely mathematical aspects of this program. It is important for us to note, however, that Husserl's teacher also acknowledged another, still deeper level of the problem of providing an adequate foundation for analysis. In his lectures, Weierstrass apparently found it necessary to give some account of the positive whole numbers themselves, the numbers which were basic to his construction of the rest of the real number system. Husserl mentions in PA that Weierstrass, after describing analysis as 'a science which is based solely on the concept of number,' proceeded to give 'an analysis of the concept of number' (PA 12 n.). And according to Husserl's record of a lecture delivered in 1880, it was with reference to 'the essence of the operation we call counting' that Weierstrass sought to explain number in the most fundamental sense.[16] Weierstrass reportedly said that in counting we begin with 'an aggregate of things or appearances,' then proceed to select from that aggregate those things which have a certain predetermined feature. We select first one thing, then another, and so on, fixing each one in our 'imagination' (*Vorstellung*) as we go. According to the report, Weierstrass then asserted that

> if we in this manner select from the relevant aggregate of things those which have the feature in question, and unite all the selected things in our imagination, we then arrive at the concept of number. Number is thus nothing other than a determinate multitude of homogeneous things.[17]

Weierstrass' assertion that numbers in the most fundamental sense are to be traced back to the mental act of counting served as the starting point for Husserl's own early research in the philosophy of mathematics. There is no evidence that Weierstrass himself saw the problem of the origin of number as a particularly difficult problem or as an issue to be taken up by philosophers, as distinct from mathematicians themselves. His student Husserl did, however, see it as a difficult problem requiring careful and detailed investigations. And when Husserl first began to speak of the 'philosophy of arithmetic,' it was this problem which he had primarily in mind. His intention was to carry through to completion Weierstrass' program for securing the foundations of analysis, and to do so by applying to the problem of the origin of number a philosophical rigor comparable to the rigor which his teacher had demanded in mathematics proper. Thus Husserl's philosophy of arithmetic took shape as an attempt to address the non-mathematical issues to which the program of arithmetizing analysis inevitably led.

Husserl's earliest work in the philosophy of arithmetic shows the deep influence of his teacher and mentor Weierstrass. But even at this early point in Husserl's development, we can find some important differences in the two men's views. And in later years, the gap between them grew even wider. The following sections of this chapter will attempt to trace the development of Husserl's views on the philosophy of arithmetic from their Weierstrassian beginnings to the mature form they attained some ten years later. As we shall see, it is possible to distinguish three major stages in this development: a first stage in which Husserl followed Weierstrass in regarding analysis as a science based directly on the concept of number; a second stage in which he considered analysis to be in essence a formal technique; and a third stage in which he characterized analysis as an instance of 'manifold theory.'

2. HUSSERL'S FIRST STAGE: ANALYSIS AS A SCIENCE OF NUMBER

In 1884, having completed a year of military service, Husserl returned to Vienna and began to devote himself to the study of philosophy, most notably with the controversial but stimulating lecturer Franz Brentano. But it was apparently not until 1886 that he applied himself to the philosophy of mathematics in particular.[18] In October of that year, on Brentano's recommendation, Husserl moved to Halle to complete his 'habilitation' under the direction of Carl Stumpf.[19] One year later – that is, in the fall of 1887 (cf. PA 8) – his habilitation essay appeared with the title: *Über den Begriff der Zahl. Psychologische Analysen.*[20] According to a note inserted by the author, the printed version of this essay was only the first chapter of a larger essay which was to have been published shortly thereafter (PA 338 n.). In fact, nothing fitting the description of this latter essay appeared – not, at least, for several years.[21] But it does nonetheless seem plausible to believe that Husserl had completed a manuscript larger than the one he actually turned over to the printer. He lectured on the philosophy of mathematics at the University of Halle during the winter semester of 1887/88, and the manuscript of these lectures goes considerably beyond the material included in the printed version of BdZ.[22] Taken together, these two texts – the printed version of BdZ and the manuscript of the 1887/88 lectures – allows us to see how Husserl, at this very early stage of his career, conceived of what he came to call the philosophy of arithmetic.

Not surprisingly, we find that Husserl's views at this time were very similar to those of his teacher Weierstrass.[23] It was apparently from Weierstrass that he had learned of the theoretical price paid for the rapid progress of modern mathematics.[24] In the introduction to BdZ, we find the following assessment of the situation:

> It is well-known what a great upsurge mathematics had made in the course of the last centuries, and also that a series of new and far-reaching tools of investigation had been discovered and that an utterly unsurveyable wealth of new knowledge had been acquired. One easily comprehends that in earlier, more creative times – times when men were still striving to fill out the great thoughts of Newton and Leibniz and to use them in making many new scientific areas fruitful – reflections on the logical nature of all the puzzling auxiliary concepts which one felt forced to introduce and apply consistently, were of necessity overlooked in favor of the striving for results, for discoveries, for exploitation of the wonderful tool. (PA 290)

Some of the auxiliary concepts Husserl has in mind here are enumerated in the following paragraph: 'the imaginary, the irrational, the differential and the integral, the continuous, etc.' (PA 291).[25] And in the manuscript, Husserl mentions negative numbers as well.[26] All these 'obscure, apparently contradictory, but in analysis still indispensable' concepts (PA 291) are, Husserl claims, in need of fundamental logical clarification.

> No reasonable person doubts the correctness of their results; no natural scientist or technician hesitates to use these same tools in the study and mastery of nature, wherever this is possible. And yet, what bad condition is the logic of the foundations of this science in! It is incredible and yet indubitable that there is not a single one of the basic concepts of analysis in regard to which clarity and insight exist.[27]

Husserl seems never to have doubted that these concepts *could* be clarified, their use theoretically 'justified'; the problem was only that, as a matter of fact, this had not yet been accomplished.

But it is not only Husserl's diagnosis of the problem that shows the influence of his teacher Weierstrass. His prescription for correcting it does so likewise. He alludes to 'the need for logical clarification, sighting (*Sichtung*) and certification of what has been achieved; for a precise analysis of the basic and intermediate concepts; for logical insight into the dependence relations of the various mathematical disciplines ...; and finally for a strictly deductive development of the whole of mathematics from as few self-evident principles as possible' (PA 291). And of course, this comprehensive need could be satisfied only if an appropriate method were followed. In the lecture manuscript Husserl clearly dismisses the possibility of founding analysis through an appeal to 'geometrical intuitions' (*geometrische Hilfsvorstellungen*).[28] At least a part of what is required, he explains, is what not only Weierstrass, but also Dedekind and Cantor had sought to provide: 'a consistent, purely arithmetical development of analysis.'[29]

Husserl remarks in a footnote in PA that

> Weierstrass used to open his epochmaking lectures on the theory of analytic functions with these statements: Pure arithmetic (or pure analysis) is a science which is based solely on the concept of number. It needs no other presuppositions, no postulates and no premises ... There then followed an analysis of the concept of number (in the sense of cardinal number). (PA 12 n.)

This summary statement of Weierstrass' position could very well stand as a statement of Husserl's own position at the time he wrote BdZ. He himself asserts in the introduction to that essay that

> it is today a matter of general agreement that a rigorous and consistent development of higher analysis ... must exclude all geometrical intuitions and begin solely with the elementary arithmetic on which it is based. But in fact this latter has its sole foundation on the concept of number – or, more precisely, in that infinitely continuable series of concepts which mathematicians call the 'positive whole numbers.' ... Thus any philosophy of mathematics must begin with an analysis of the concept of number. (PA 294f.)

This early Weierstrassian view of analysis as a science of number was later abandoned by Husserl, as we shall see in the section that follows. But for the first few years of his work on the philosophy of mathematics, from 1886 until late 1889 or early 1890, it was the idea which guided all of his more specific investigations.

Husserl's early conception of the philosophy of arithmetic can scarcely be understood unless seen in the light of Weierstrass' influence. It would be incorrect, however, to conclude that Husserl remained wholly within the parameters set up by his teacher. For it seems that there were significant differences between them even at the early stage of Husserl's development now under consideration. It is not unfair, in fact, to say that Husserl, from the very beginning of his work in this area, sought a foundation for analysis yet more radical and more rigorous than that which Weierstrass himself had envisioned.[30]

In apparent contrast to his teacher, Husserl maintained that inquiry into the foundations of mathematics was not strictly speaking a *mathematical* task at all. As he put it in his lecture manuscript, 'not every scientific concern with mathematics is a mathematical concern.'[31] The most fundamental questions regarding mathematics are in fact *philosophical* questions for which mathematical expertise alone does not suffice. Indeed, it is possible that even a 'great mathematician' would be guilty of the 'weightiest errors' on philosophical questions regarding the foundations of his discipline. So different are the two types of inquiry that for the mathematician *qua* mathematician – or for 'the mathematician who is a mathematician and nothing more' – these fundamental questions are virtually irrelevant. It is only, or at least chiefly, because a mathematician is in fact 'something more than a mere machine for the calculative derivation of theorems for accepted principles' that he ought to be concerned with such questions.

> This limitation to ever more specialized fields [that is characteristic of modern science in general] is nothing that constitutes value or worth. It is only a necessary evil. The complete researcher who strives to be a complete human being as well should never lose sight of the relation of his science to the more general and higher epistemic goals of mankind. Professional restriction to a single field is necessary; but it is reproachable to become fully absorbed in such a field. And he must appear especially reproachable who is indifferent even to the more general questions which concern the foundations of his science, as well as its value and its place in the realm of human knowledge in general.[32]

It might of course happen that the mathematician's mathematical work would be adversely affected by a failure to think philosophically about his activity (cf. PA 290f.). But it is the mathematician's own sense of scientific self-responsibility, not the needs of theoretical praxis, which forms the most important motivation for the inquiry into the foundations of the discipline.

Given this insistence on the need for philosophical investigations into the foundations of mathematics, we are led to ask just what Husserl meant by 'philosophy' at this stage of his career. Although this question is not addressed systematically in any of Husserl's early writings, it is nonetheless clear that he regarded 'psychology' as a very important part of philosophy, perhaps the most important part.[33] In BdZ, Husserl assigns the fundamental problem of analyzing the number concept not just to philosophy in general, but more particularly to 'psychology.' The 'tools' (*Hilfsmittel*) to be used in the analysis of this concept 'belong to psychology,' and furthermore they 'must do so if such an investigation is to arrive at assured results' (PA 295).

It is important to understand just what Husserl meant by 'psychology' in this context. In a seminal work entitled *Psychologie vom empirischen Standpunkte* (1874), Husserl's teacher Brentano had distinguished between the inductive, explanatory study of consciousness (which he called 'genetic psychology') and the purely descriptive analysis of the phenomena of consciousness (which he later came to call 'descriptive psychology').[35] Though Brentano had at first viewed descriptive psychology merely as a preliminary part of genetic psychology, by the 1880s he had come to assign a considerably more significant role to this discipline. He regarded it as an *a priori* science in its own right and, furthermore, as the discipline charged with establishing the foundations of other *a priori* sciences.[36] Husserl's own early views were similar to those of Brentano. Thus when he advocated the use of psychology in the study of the number concept fundamental to analysis, he undoubtedly had the descriptive psychology outlined by Brentano in mind.[37]

In any case, Husserl clearly did not wish to assert that numbers were peculiarly subjective, mental entities, or that analysis was in some sense a branch of empirical psychology. In the introduction to BdZ, we find Husserl raising the following issue:

> Admittedly one might ask at first: what does number have to do with psychology? To this question we would pose another: what do space, time, color, intensity, etc., have to do with psychology? Is space not the object of the

> geometer, color the object of the physicist, and so on?
> And yet these concepts have given rise to an extensive psychological literature that is growing daily. (PA 295)

This answer is somewhat cryptic, to be sure; it does, however, suggest that the early Husserl saw no contradiction between the view that number is in some sense an 'object' and the view that the concept of number is to be analyzed by psychology. Number as such, it seems, is no more subjective or psychological than space, time, color and intensity. And the lecture manuscript makes it quite clear that Husserl did not consider mathematics a branch of inductive, empirical science. He claims in this manuscript that there is an 'essential difference' (*wesentlicher Unterschied*) between the 'objects' (*Gegenstände*) of mathematics and the objects of the inductive natural sciences:

> Mathematics deals always and everywhere only with purely logical dependencies of magnitudes and positional relationships, the natural sciences with the real relationships of things to each other (the former immediately given, the latter hypothetical and inferred). Thus every mathematical law is subject to the principle of contradiction, in the sense that whoever denies such a law is asserting an absurdity. A natural law is not subject in this way to the principle of contradiction, but only to the precepts of induction. It is an inductive law, and whoever denies it merely slips into an assertion that is highly – or, in the optimal case, infinitely – improbable.[38]

It is true that at a later stage in his career, Husserl temporarily advocated an interpretation of analysis which minimized the difference between this discipline and the inductive natural sciences. But at the early stage now under consideration, Husserl had not yet fallen into the view which he was ultimately to reject as 'psychologistic.'[39]

From the above we can see that Husserl's characterization of the initial problem in the philosophy of mathematics departs from Weierstrass' views in two significant respects. First, Husserl describes this problem as a distinctly non-mathematical, philosophical one. Second, Husserl finds it necessary to use the 'tools' of descriptive psychology in attacking the problem; indeed, he claims that the problem is an inherently 'psychological' one. Of course, neither of these positions stands in direct contradiction to what Weierstrass himself had maintained. It could be argued, for example, that Weierstrass' analysis of the number concept is itself implicitly psychological.[40] Nonetheless it is clear that Husserl, from his own philosophical or psychological point of view, finds the number concept considerably more problematic than had Weierstrass. Where Weierstrass seems to have dealt with the question in a few sentences, Husserl wrote an essay of at least sixty-four pages on the subject.

More significant perhaps are the actual differences of content between the two analyses of the number concept. Husserl might have agreed with Weierstrass that we discover the origin of number, and thereby clarify the number concept, by examining the mental act of 'counting' (*Zählen*). But what is counting? Husserl's analysis of this phenomenon is by no means identical with that of Weierstrass. Many particular differences will emerge later in the course of the present study, but one of them must be mentioned even at this early stage.

It appears that Husserl, from the very beginning of his work in the philosophy of mathematics, attached great importance to the distinction he drew between the 'authentic' (*eigentliche*) and the 'symbolic' presentation of number (cf. PA 6f., 190ff.). There is no evidence that Weierstrass made this or any similar distinction. In fact, Husserl later wrote that it was his teacher Brentano who had first called attention to the issue. After introducing the distinction in the text of PA, Husserl inserts the following note:

> I am indebted to [Brentano] for a deeper understanding of the preeminent significance of inauthentic presentation for our entire mental life; this is something which, so far as I can see, no one before him had grasped.[41]

Now it is true that in BdZ, which Husserl wrote several years prior to the publication of PA, the distinction between authentic and symbolic presentation does not play a central role. Yet this important distinction is at least mentioned in the early essay. At one point Husserl alludes to the fact that 'we achieve authentic presentations only of the very first numbers, while we can think larger numbers only symbolically, only by an indirect route, so to speak' (PA 322). That Husserl attached importance to this distinction is shown by the fact that one of the eight theses he chose to defend in a formal disputation in the summer of 1887 was the following: 'In the authentic sense one can barely count beyond three' (PA 339). But the clearest evidence for Husserl's early use of this notion is the manuscript of his 1887/88 lectures. Here we find included what could plainly serve as a preliminary draft of Part Two of Husserl's later work *Philosophie der Arithmetik* (1891).[42] Many of the issues discussed at length in this part of the later work are given at least brief consideration in the corresponding sections of the manuscript. We find, for example, a discussion of the distinction between symbolic and authentic presentation, as well as the claim that all of our number concepts, except for the first few in the number series, are and can only be 'symbolic concepts.'[43]

We cannot discuss the distinction between authentic and symbolic presentation in detail here; we shall return to it later in the course of this study.[44] We must note nonetheless that the early Husserl seems to have regarded it as the very cornerstone of his work in the philosophy of mathematics. As he was to write later in *Philosophie der Arithmetik*, the fact that almost all of our number presentations are symbolic rather than authentic 'completely determines the character, sense and purpose of arithmetic.'[45] It is only when we realize that our number concept is almost wholly 'symbolic' in nature that we can begin to understand the specific character of even the most elementary arithmetical knowledge.

The early Husserl apparently thought that this insight could in some way help to clarify the expanded number concept that is central to analysis and thus bring to completion the philosophical project he had begun under the inspiration of Weierstrass. Unfortunately, however, both the lecture manuscript and BdZ itself end before this latter issue is taken up. One can only conjecture about Husserl's reasoning here. Perhaps his view was simply this: Since even our most elementary number concept is largely 'symbolic,' there is no intrinsic mystery regarding the introduction of other 'symbolic' concepts, such as those of negative, rational, irrational and imaginary numbers. The original or 'authentic' number concept has already been broadened to include numbers not actually given to us, so why should we not broaden it further? We are perfectly justified in taking this step, provided of course that we are not led into actual errors or contradictions by so doing. The 'puzzling auxiliary concepts' of analysis are thus, in the end, no more mysterious than the number 22.

This may or may not represent a correct interpretation of Husserl's views in 1887.[46] It does go somewhat further than Weierstrass had in suggesting a connection of sorts between our ordinary experience of number and the obscure and puzzling number concepts used in analysis. But whatever Husserl's early thoughts on this subject may have been, it is certain that he did not long adhere to them. Shortly after the publication of BdZ, Husserl moved forward to a substantially different view of analysis – one which marked a much sharper break with the views of Weierstrass than anything we have observed up to this point.

3. HUSSERL'S SECOND STAGE: ANALYSIS AS A FORMAL TECHNIQUE

Students of Husserl's philosophy sometimes argue that the most significant development in his thinking occurred around the year 1900, or shortly thereafter. This may be correct, but in regard to the specific issues with which he was concerned in his philosophy of arithmetic, we find that the most important developments occurred long before this time. It is with the first of these – the development that led to what we call the second stage of Husserl's philosophy of arithmetic – that this section is concerned. At some time after the publication of BdZ in the fall of 1887, but certainly no later than February, 1890, Husserl came to espouse a fundamentally new position regarding the character of analysis.[47] He found himself forced to abandon some basic assumptions which he had inherited from Weierstrass – assumptions which had guided all of his own work in the philosophy of mathematics up to that time. He was led to formulate the crucial philosophical problems regarding analysis in a new manner. These newly formulated problems, in turn, became the basis for a period of intensive reflection and research on the foundations of mathematics in the early 1890s.

In 1891, while Husserl was still a *Privatdozent* at the University of Halle, he published Volume I of a planned two-volume work with the titel: *Philosophie der Arithmetik. Logische und psychologische Untersuchungen.*[48] This volume has long been regarded by scholars as the most important document pertaining to Husserl's early views on the philosophy of mathematics. However, the highly problematic character of the work has been largely overlooked.[49] In fact, this volume poses some extremely difficult questions of interpretation, questions which have almost never been addressed or even acknowledged.[50] The fundamental problem is the following: By 1891, when PA was published, Husserl had already arrived at the new view of analysis alluded to in the preceding paragraph. Indeed, we find statements plainly based on this view in the Introduction to Part One of the work, in the final chapter of Part Two, and especially in the Preface to the work as a whole. And yet the great bulk of the work nonetheless derives from the year 1887 – that is, from the period *before* Husserl had arrived at his new view of the fundamental character of analysis. As Husserl himself points out in the Preface, 'some of the psychological investigations of the present volume were contained almost verbatim in my habilitation essay, from which a booklet of four signatures was printed in the fall of 1887 under the title *Über den Begriff der Zahl ...*' (PA 8).[51] The result of this situation is that we find Husserl speaking in PA from two significantly different points of view: the early view of the nature of analysis which he had taken over from Weierstrass and the later view which we have yet to examine. It is no surprise, therefore, that some fifteen years after the publication of PA Husserl expressed regret about the work in the following terms:

> It was not without reason that my conscience bothered me at the publication [of PA]. Actually I was already beyond it when I published it. After all, it derived in its essentials from the years 86/87.[52]

The two points of view expressed in PA are not necessarily incompatible in all respects. Nonetheless it is essential, if that work is to be understood properly, that they be distinguished. Let us look more closely therefore at the new view of analysis to which Husserl advanced in the period after the publication of BdZ.

In 1887, at the time BdZ was written, Husserl had viewed analysis or the *arithmetica universalis* as a science founded ultimately on the concept of number alone, as we saw in the preceding section. All of the 'puzzling auxiliary concepts' utilized in analysis, according to the view he had held at this time, 'have their origin and basis in the elementary number concepts and the relations which connect them.'[53] This view of analysis as a 'science' of number and number-relations – a view suggested by Weierstrass' lectures on mathematics – had been the motive for Husserl's own philosophical study of the concept of cardinal number. By showing that even this elementary concept was largely 'symbolic' in nature, he felt he could provide a foundation for the more complex and puzzling number concepts of higher mathematics.

After the publication of BdZ, however, Husserl at some point came to realize that this was not in fact an adequate view of the matter. He found himself compelled to abandon the thesis that analysis was a science of number, and likewise the thesis that the auxiliary number concepts have their 'origin and basis' in the elementary number concept alone. We see evidence of this in the Introduction to Part One of PA.[54] Even though PA is almost entirely devoted to studying the concept of cardinal number, Husserl nonetheless refrains from asserting outright that this concept is the basis of analysis. He does mention the opinion that cardinal number is the basic concept of analysis, commenting as he does so that this opinion is one shared by 'many' theorists – 'including very important mathematicians, such as Weierstrass' (PA 11f.). But he also points out that other theorists hold different views: some regard ordinal number as the fundamental concept, still others take linear magnitude as basic (PA 12). Husserl himself takes only a 'preliminary' stand on the issue. He does tentatively associate himself with the view of Weierstrass, since it is, he says, 'the obvious view' (*die zunächst liegende Ansicht*) (PA 12). But he adds immediately that this preliminary stand 'will not in any way prejudice our final decision.' In fact, he says, the results of the intended second volume of PA might well show the 'untenability' (*Undurchführbarkeit*) of Weierstrass' view (PA 12).[55]

There is a straightforward explanation for the curiously awkward character of these remarks.[56] By the time Husserl wrote the Introduction (presumably in 1890 or 1891), he had *already* come to realize that analysis was in fact *not* based on the concept of number, as Weierstrass had taught and as he himself had initially assumed. Despite the fact that he had decided to go ahead with the publication of his earlier study in its original form, he no longer regarded the number concept as the sole basis for analysis as a whole. He himself says as much in the Preface to PA (which is dated April, 1891). In outlining the material to be treated in the work, he draws a sharp distinction between the 'arithmetic of number' (*Anzahlenarithmetik*) and 'universal arithmetic' (*allgemeine Arithmetik* (PA 7). The former is clearly based on the concept of number; the latter, however, is not. It is *not* the case, Husserl indicates, that the region of cardinal number is what 'universal arithmetic in the primary and original sense masters,' nor is it the concept of cardinal number that 'mediates the application' of universal arithmetic in all instances (PA 7). Husserl speaks in a similar vein in the *Selbstanzeige* (author's announcement) for PA, also published in 1891. In this announcement he mentions that the symbolic methods of universal arithmetic are 'quite different in kind' from those of what he here calls 'numerical arithmetic' (*numerische Arithmetik*) (PA 287f.). Thus it is clear that by 1891, Husserl had come to see that the conceptual and methodological foundations of analysis were not what he had originally thought them to be.

A letter written by Husserl to Carl Stumpf in February, 1890, confirms this interpretation and shows, in addition, that he had arrived at his new view of analysis well before 1891.

> The opinion [Husserl writes] which still guided me in working out the inaugural dissertation [BdZ], that the concept of number formed the basis of universal arithmetic, soon proved to be false. (The analysis of the ordinal numbers already led me to this conclusion.) No technique, no 'imaginary representations,' can lead one to deduce the negative, rational, irrational and the various complex numbers from the concept of number ... The fact is that 'universal arithmetic' (including analysis, theory of functions, etc.) finds *application* to numbers ('theory of numbers') ...[57]

This letter indicates in unmistakable terms that by 1890 Husserl had moved a great distance from the views with which he had begun his work in the philosophy of mathematics. He had come to see quite clearly that no study of the number concept alone would suffice to provide an ultimate philosophical foundation for the puzzling concepts of higher mathematics. Not even the distinction between 'authentic' and 'symbolic' presentation of number would be sufficient. The larger numbers might in some sense be 'symbolic' concepts; but they are not symbolic, Husserl seems to imply, in the manner in which a negative or irrational number is. We cannot provide an ultimate foundation or justification for the concepts introduced in analysis simply by studying the concept of number.

At this point we must ask what analysis *is*, according to the Husserl of the early 1890s, if it is not a science of numbers and their relations. Did he perhaps hold that some other concept, such as ordinal number of linear magnitude, was the basis of this discipline? It seems that he did not. Though it is true, Husserl writes in the letter to Stumpf, that the negatives, rationals, irrationals, etc., cannot be derived from the concept of *cardinal* number, 'the same holds for concepts of ordinal numbers, concepts of magnitude, etc.'[58] In fact, as he puts it in the Preface to PA, 'there is by no means one *single* kind of concept, whether it be cardinal number, ordinal number or something else, which *everywhere* mediates the application' of universal arithmetic (PA 7).

How then *is* this discipline to be understood? In his letter to Stumpf Husserl answers this question as follows:

> ... the *arithmetica universalis* is not a science but a piece of formal logic ...[59]

Analysis, in other words, is not a science of number in the way elementary arithmetic is. Its methods have more in common with the methods of logic than they have with those of elementary arithmetic. But to understand just what Husserl meant by this important assertion – an assertion which stands at the center of the second stage in the development of his philosophy of arithmetic – we must briefly examine the views which Husserl held at this time on the nature of logic in general, and of formal logic in particular.

In the Introduction to BdZ (1887), Husserl had mentioned in passing that logic is a 'practical discipline,' more specifically 'a technology of correct judging' (*Kunstlehre des richtigen Urteilens*) (PA 291). Although he did not explain what he meant by a 'technology of correct judging,' there can be but little doubt that he had derived this view of logic from his teacher Brentano.[60] While in Vienna Husserl had attended Brentano's lectures on logic in the winter semester of 1884/85. And it is known that in these lectures Brentano defended the thesis that logic is the 'art of correct judgment' (*Kunst des richtigen Urteils*).[61] By this Brentano had meant that logic is not a theoretical science, but rather a practical discipline whose task is to prescribe rules or norms for the attainment of knowledge.[62]

This background helps us to understand what Husserl meant in the 1890 letter to Stumpf by 'formal logic.' He gives only a very brief elucidation of the term in the letter itself: formal logic can be defined, he says, as a 'technique for manipulating signs (*Kunst der Zeichen*).'[63] Fortunately, this conception of formal logic is developed at much greater length in a manuscript written in March, 1890, just one month after the letter to Stumpf.[64] This manuscript, which was evidently a preliminary study for a planned appendix to the second volume of PA (cf. PA 6), is entitled 'Zur Logik der Zeichen (Semiotik).' The manuscript clearly demonstrates that at this time Husserl's views on the nature of logic were still essentially those he had acquired from Brentano. Indeed, we find a reference in the manuscript to 'Brentano's epoch-making investigations' in the field of logic (PA 346). It is true that Husserl's definition of logic had changed slightly. He now defines it not as the 'technology of correct judging,' but rather as the 'art of knowledge' (*Kunst der Erkenntnis*) (PA 373). But this seems to reflect only an expansion of the Brentanian conception of logic, not a fundamental departure from it. Husserl still claims that

> the task of logic is ... *to master the natural procedures of the judging mind*, to test them, to acquire insight into their epistemic value, in order finally to be able to determine their limits, scope and bearing *and set up general rules concerning them.*[65]

Logic in general, then, is a practical discipline concerned with evaluating the procedures of the 'judging mind.' And what is 'formal logic' in particular, according to the Husserl of 1890? In the manuscript formal logic is again described as a 'logic of signs' (PA 373). What this means, we learn, is that formal logic is a subdivision of logic in general concerned specifically with 'algorithmic' procedures. The task of formal logic is 'above all to clarify the algorithmic procedures that have to such a great extent become the vehicle of progress in the exact sciences and *to set up rules for such methods*' (PA 365). In other words, what logic as such is to the judging mind in general, formal logic is to the mind that forms judgements with the help of symbolic methods.

Now given that Husserl adhered in 1890 to this conception of the nature of formal logic, we may infer what his conception of analysis or the *arithmetica universalis* must have been at this time. Since he stated that analysis was to be regarded as a 'piece of formal logic,' he must have viewed it as at bottom a practical discipline whose task was to set up rules for the correct derivation of symbolic judgments. According to this view, analysis is not a theoretical science, but rather an 'art' (*Kunst*), a 'technology' (*Kunstlehre*), or an 'algorithm' (cf. PA 7).

Once he had come to the recognition that analysis is not simply an extension of elementary arithmetic, but rather an enterprise of an entirely different kind, Husserl felt himself compelled to expand his view of the problems to be addressed in the philosophy of arithmetic. This expansion is reflected in his statement of the topics to be covered in the two volumes which he planned for PA. In Volume I (the only volume ever published), Husserl deals mainly with 'psychological questions' pertaining to the authentic and symbolic presentation of number (cf. PA 6f.). This volume, in other words, is a revised and expanded version of his habilitation essay BdZ. For, as Husserl somewhat cryptically observes in the Introduction to PA, his psychological studies of the number concept 'would in no way be robbed of their value' even if it should turn out that universal arithmetic were in fact not based on this concept (PA 12). Number is one of several 'fundamental concepts of human knowledge in general,' and thus it can lay claim to a certain interest in its own right (PA 13).

But it is in Husserl's plan for Volume II of PA that we see the clearest reflection of his new understanding of the basic character of analysis. A whole new set of philosophical problems has appeared, problems quite different from those suggested by Husserl's former mentor Weierstrass. One such problem is that of providing a 'full logical clarification of the true sense of universal arithmetic' (PA 7). Under this heading Husserl presumably meant to present and defend his view that analysis is in essence not a science of number but a formal technique, an 'algorithm ... which masters a series of quite distinct conceptual regions' (PA 7). Another issue which Husserl promised to address in the second volume of PA was that of providing a 'justification for the calculative utilization of the quasi-numbers arising from the inverse operations (the negative, imaginary, fractional and irrational numbers)' (PA 7). It is significant that Husserl now implies a sharp distinction between these 'quasi-numbers' and the larger positive whole numbers. Even though the latter are themselves 'symbolic concepts,' they do not require the special justification that the quasi-numbers do.

Although Husserl worked on the planned second volume of PA for a time, he was never able to bring it to completion. His failure to complete it was undoubtedly due at least in part to the fact that he soon came to find a further important modification of his views necessary. This modification, which led to what we may call the third stage in the development of Husserl's philosophy of arithmetic, is the subject of the following section.

4. HUSSERL'S THIRD STAGE: ANALYSIS AS MANIFOLD THEORY

Husserl worked on the second volume of PA, or on problems closely related to the announced content of that volume, at least until the year 1894. In fact, it seems that he still had the intention of completing and publishing the volume as late as November, 1894.[66] But in time his central interest shifted from the philosophy of arithmetic to the philosophy of logic in general.[67] As a result of his work in this latter area, he eventually became convinced that his earlier, Brentanian view of logic as an essentially practical discipline, as a 'technology of correct judging,' had been seriously in error. Instead, he came to maintain, logic is a pure, *a priori* science whose primary concern is not the judging mind, but rather the realm of unchanging, ideal meanings. This new view of logic is of considerable interest in its own right. But what is important in the present context is the effect which it had on Husserl's 'philosophy of arithmetic.' For although this area was no longer his chief philosophical interest, he did at least continue to reflect on the nature of analysis and to draw the appropriate implications from his new philosophy of logic. As we shall see, he could not remain content with the view that analysis is merely an algorithm with the aid of which the mind is able to draw correct symbolic judgments.

Because of his changing philosophical interests, his attempt to probe carefully and systematically into ever more fundamental problems, Husserl never wrote a book-length work devoted exclusively to the exposition of his new understanding of analysis. The books he did publish, however, allow us to reconstruct at least the broad outlines of the third stage of Husserl's philosophy of arithmetic. In the Prolegomena to his *Logische Untersuchungen* (1900/1901),[68] we find several sections in which Husserl's newly formulated views on the nature of analysis are stated briefly but directly. The later work *Ideen zu einer reinen Phänomenologie und phänomenologischen Philosophie* (1913)[69] touches on these issues only in passing, but *Formale und transzendentale Logik* (1929)[70] and *Die Krisis der europäischen Wissenschaften* (1936)[71] contain somewhat more extensive discussions. In addition, a number of unpublished or posthumously published manuscripts offer a more detailed picture of Husserl's later answers to the questions he had raised at the very beginning of his career. The most important of these is the partially preserved manuscript of the so-called '*Doppelvortrag*' (double lecture) which Husserl delivered to the Göttingen Mathematical Society in November and December of 1901.[72] Other manuscripts, such as that pertaining to Husserl's 1906/07 course on logic,[73] also shed light on his later views.

These sources indicate that what we have called the third stage in the development of Husserl's philosophy of arithmetic was in fact its final stage. Husserl's conception of the basic character of analysis did not undergo further major change after the publication of LU, nor did his understanding of the philosophical tasks associated with the foundations of that discipline. In this study we shall attempt a full exposition of the mature and final version of Husserl's philosophy of arithmetic. It will be helpful, however, for us to examine some of its major characteristics here, before we begin our more detailed exposition.

According to the mature Husserl, analysis is neither a science of number nor a formal algorithm, but rather a theoretical discipline directed toward a distinctive type of object region. It is, to use his technical term, an instance of 'manifold theory.' In the theory of deductive systems which Husserl worked out in the 1890s and presented to the public in LU and FTL, the term 'manifold' refers to 'a region (*Gebiet*) which is uniquely and solely determined by falling under a theory of [a given] form.'[74] A manifold, in other words, is not merely a set of objects; it is not something which we might encounter prior to constructing the formal deductive system that is correlated with it. It is rather a region of objects defined solely by the relationships set up for them in a formal deductive system, and thus a region to which we have access only through the formal system in question. According to the Husserl of LU and FTL, the 'real number system' central to analysis is in fact a mathematical manifold, an object region defined by a formal deductive system.[75] 'Real numbers' are objects in a sense, but they cannot be encountered in the way the numbers studied in elementary arithmetic can. They are not mere fictions void of all reality, but they are entities of a very different character from the numbers presented to us in the act of counting.[76]

In one sense, Husserl's mature interpretation of analysis as an instance of manifold theory has something in common with his early view of analysis as a science of number. In both cases, analysis is viewed as a theoretical discipline concerned with a region of objects, not as an essentially practical discipline or algorithm. But of course, Husserl's mature position is by no means a reversion to the position he had taken early in his career under the influence of Weierstrass. He remarks in LU that 'only if one is ignorant of the modern science of mathematics, particularly of formal mathematics, and measures it by the standards of Euclid and Adam Riese, can one remain stuck in the common prejudice that the essence of mathematics lies in number and quantity' (LU I 252f. [244]). In opposition to this outdated view, Husserl places 'what has been developed as 'pure analysis' or, better, as *formal* mathematics,' in 'the field of pure or formal logic' (LU I 252 [243f.]). Thus we see that Husserl regards analysis as a part of formal logic in this third stage of his development, just as he had in his second. What has changed is his conception of logic itself. Having abandoned his Brentanian view of logic as an art of forming correct judgment, he now regards logic – or at least the 'pure logic' which is its core – as a genuinely theoretical discipline. The formal logic of deductive systems, in particular, is not a mere 'technique for manipulating signs,' as Husserl had defined it in his 1890 letter to Stumpf. According to his mature position, its concern is with truths about a certain region of objects rather than with techniques to aid the judging mind.

Had Husserl chosen to write a book-length work devoted to the exposition of his 'philosophy of arithmetic' in its mature form, the work's basic structure might well have been similar to that planned for PA. Despite the essentially new conception of analysis, Husserl's understanding of the philosopher's role in investigating the foundations of that discipline remained largely unchanged. He continued, for example, to draw a sharp distinction between mathematical and philosophical tasks:

> The nature of the case really demands a thorough-going *division of labour* here. The construction of theories, the strict, methodical solution of all formal problems, will always remain the home domain of the mathematician. (LU I 252 [244])

The philosopher, on the other hand, though he does not seek to 'meddle in the work of the specialist,' is concerned nonetheless to 'achieve insight in regard to the sense and essence' of the latter's achievements (LU I 254 [245]). 'The interest of pure theory' alone motivates his inquiry, and thus he 'is not content with the fact that we find our way about in the world, that we have legal formulae which enable us to predict the future course of things ...' (LU I 254 [254]). The mathematician, by contrast, 'is not really the pure theoretician, but only the ingenious technician, the constructor, as it were, who, looking merely for formal interconnections, builds up his theory like a technical work of art' (LU I 253 [244]). Hence sustained philosophical reflection and research is essential if the mathematician's technical achievement is to become 'pure, genuine, theoretical knowledge' (LU I 254 [245]).

It was not only this general understanding of the philosopher's role which remained unchanged in the transition to Husserl's third stage, but also his view of the specific problems to be addressed. The problems Husserl had planned to take up in Volume II of PA, for example, clearly remained important to him at the later stage. He continued to see the problem of clarifying the 'sense' of analysis as a major philosophical task, though of course he had come to view analysis as manifold theory rather than a formal technique or algorithm. But he also continued to concern himself with the problem of justifying the introduction of the so-called 'quasi-numbers.' In fact, what Husserl came to call 'the problem of the imaginary'[77] was apparently the central theme of the *Doppelvortrag* he delivered in 1901. He asks in the manuscript of that lecture:

> With what right may something absurd (*das Widersinnige*) be processed calculatively? With what right can something absurd be applied in deductive thinking as though it were something coherent (*Einstimmiges*)? How are we to explain that one can operate with something absurd according to rules, and that if the absurdity is dropped from the propositions, the resulting propositions are correct? (PA 433)

This statement of the problem is obviously similar to that given in the Preface to PA, and it seems very likely that Husserl's solution to the problem was also substantially the same as that which he had envisioned earlier in his career.[78] We shall examine this solution in some detail later in the course of this study.

But what of the problems addressed in Volume I of PA? Do the 'psychological questions' regarding the authentic and symbolic presentation of number retain their importance in the context of Husserl's mature philosophy? To judge from Husserl's later comments about this volume, it would certainly seem that they do. In LU he lists number (*Anzahl*) among the 'objective categories' whose respective 'origins' are to be investigated, and whose 'essences' are to be clarified, by philosophy (LU I 244 [237f.]). He reports moreover that he continues to find his earlier psychological study of number 'clear and instructive' (LU I vii [42]). In FTL likewise, Husserl writes, with reference to such disciplines as elementary arithmetic (*Anzahlenarithmetik*), that 'one can never spare ... bringing out their *fundamental concepts* ... and the *actual axioms* relating to these concepts' (FTL 104 [99]).[79] And once again he speaks approvingly of his early work on this problem: 'In spite of its immaturity as a first book, [PA] presented an initial attempt to go back to the spontaneous activities of collecting and counting, in which collections ... and cardnal numbers are given in the manner of something that is being generated originaliter, and thereby to gain clarity respecting the proper, the authentic, sense of the concepts fundamental to the theory of sets and the theory of cardinal numbers' (FTL 90f. [86f.]).

These quotations from LU and FTL plainly suggest a deepseated continuity in Husserl's approach to the problem he had addressed first in BdZ and PA. Even in the third and final stage of his philosophico-mathematical development, it seems, Husserl continued to regard the origin of number as an important philosophical problem. What is more, he also continued to approve of the method he had employed in dealing with this problem – or so it seems, at least. But the issue of continuity in Husserl's work on the origin of number is not quite so clear-cut as it might appear at this point. For in fact, the mature Husserl later directed sharp criticism at his own early work in the philosophy of arithmetic. He came to admit that he had been guilty early in his career of a fundamental error in the philosophy of mathematics, the error of 'psychologism.' This admission is usually understood by Husserl scholars to refer to the analysis of number presented in the published volume of PA. But in order to determine whether this common interpretation of Husserl's later self-criticism is correct, we must examine it and its context somewhat more closely.

5. THE PROBLEM OF PSYCHOLOGISM IN HUSSERL'S EARLY WRITINGS

We have seen that the early Husserl sought to improve upon his teacher Weierstrass' account of the origin of number by employing the descriptive psychological method of Franz Brentano. The study of number presented in BdZ and PA was thus described in the subtitles of those works as a 'psychological' one.[80] In the works themselves, it was further characterized as a study of our 'presentation' (*Vorstellung*) of number and of the psychological origin of our number concept.[81] Now in a well-known review of PA which appeared in 1894, Gottlob Frege directed harsh criticism at this attempt on Husserl's part to found mathematics through 'psychology.' Frege argued that Husserl had failed to recognize the distinction between number and the presentation of number. According to Frege, this failure had thoroughly vitiated the study of the concept of number presented in PA. For, he argued, 'if my presentation of number is not itself the number, then the basis of the psychological approach has been removed, insofar as the latter aims at investigating the essence of number.'[82]

In the Prolegomena to LU (1900), Husserl seems at least to express agreement with Frege's criticism of his own earlier work. In PA, Husserl had written that he could not share Frege's views regarding the necessity of founding mathematics on logic rather than psychology (PA 118f.). In the Prolegomena, however – a work in which Husserl himself presents a sustained and detailed critique of what he calls 'psychologism' in the field of logic – we find an acknowledgement of an important change in the author's assessment of Frege. In a footnote, Husserl remarks with explicit reference to the passage in PA just mentioned: 'I need hardly say that I no longer approve of my own fundamental criticisms of Frege's antipsychologistic position set forth in my *Philosophie der Arithmetik* ...' (LU I 169 n. [179 n.]). Elsewhere in the first volume of LU, Husserl comments even more directly on the development of his philosophical thought. The Foreword, for example, ends with the following often-cited remark:

> As regards my frank critique of psychologistic logic and epistemology, I have but to recall Goethe's saying: There is nothing to which one is more severe than the errors that one has just abandoned. (LU I viii [43])

And most directly of all, Husserl states unequivocally, in a footnote to Section 57 of the Prolegomena, that he has now moved beyond his own earlier 'psychologistic view of the philosophy of mathematics' (LU I 212 n. [213]).

These comments have been interpreted by most students of Husserl's development to refer to the latter's earlier 'psychological' study of the concept of number.[83] Husserl is understood to be admitting that by studying the presentation of number and seeking the 'psychological origin' of the concept of number, he had in effect been guilty of a decisive and fundamental error. Despite his favorable references to the earlier works in FTL and even in LU itself,[84] in other words, Husserl is taken to be rejecting as psychologistic the basic methodological assumption which had underlain his study of the number concept in BdZ and the published volume of PA.

If this interpretation of Husserl's self-critical remarks were correct, it would obviously force us to raise serious questions about the status of his early writings from the point of view of his mature philosophy of arithmetic. Even if the central *problems* addressed in those writing remain relevant in the context of his mature position, the *solutions* offered would have to be rejected. But as we shall see, the interpretation in question is not in fact correct. When Husserl referred in LU to his own earlier 'psychologistic view of the philosophy of mathematics,' he was not referring to the study of number presented in BdZ and PA.[85]

The common misinterpretation of Husserl's remarks rests on a failure to view his early works in the context of the development of his philosophy of mathematics as a whole. Husserl's ultimate concern was always to clarify the achievements of modern mathematical analysis in its entirety. He wanted to contribute something, as he says in the preface to PA, to the 'true philosophy of the calculus' (PA 7). As we have seen, he had begun with the view, taken over from Weierstrass, that analysis as a whole could be interpreted as the science of numbers and their relations. Subsequently, however, he had rejected this assumption in favor of the view that 'the *arithmetica universalis* is not a science but a piece of formal logic.' Given the Brentanian view of logic which he held at the time, this assertion implied for him that analysis was in some sense a practical discipline rather than a theoretical one. It was, as he had put it, an 'algorithm' devised to aid the mind in arriving at certain symbolic propositions.

Now there can be but little doubt that Husserl later came to regard his earlier, Brentanian view of logic as a form of psychologism. The view that logic is essentially a practical discipline, an 'art' or 'technology' of correct judging, is subjected to extensive criticism in the Prolegomena.[86] And Husserl was clearly aware that in criticizing this view, he was criticizing the position which he himself had taken earlier in his career under Brentano's influence.[87] This criticism implied in turn a correlative judgment of his earlier views on the nature of mathematics, or at least those which had emerged in what we have called the second stage of his development. For since Husserl had regarded analysis as a part of logic, any fundamental misconception of the latter must have resulted in an equally fundamental misconception of the former. He could not have been guilty of psychologizing logic, in other words, without also having psychologized analysis.

We suggest that it is in this sense and this sense only that the early Husserl's philosophy of mathematics is tainted by psychologism, as that term is used in LU.[88] Husserl's subsequent recognition that his early work on the philosophy of mathematics had been guilty of psychologism, in other words, should not be taken as a rejection of the descriptive-psychological method used in his early writings. Nor should this admission be taken to imply that he had in some way 'psychologized' the concept of number in BdZ or PA. His early psychologism in fact had nothing to do with his use of a psychological method or with his treatment of the concept of number. Rather, it consisted in his having held, on the one hand, that analysis was a part of logic, while holding, on the other, that logic was an essentially practical discipline.[89]

Clearly, if Husserl *had* in his early writings treated number as a subjective phenomenon, he would have been guilty of a serious form of psychologism. But it is quite possible, as Husserl points out in FTL (86 [82]), to subjectivize or psychologize *logical* entities such as judgments and inferences while recognizing that *mathematical* entities like numbers have an objective, non-mental status. The difference between a judgment (or proposition) and an act of judging is not as obvious as the difference between a number and an act of counting. Indeed, Husserl argues in the Prolegomena that, although some people make the mistake of reducing judgements and inferences to the mental acts of judging, inferring, etc., 'no one' makes the analogous mistake in regard to numbers.

> No one regards the theories of pure mathematics, e.g. the pure theory of numbers (*Anzahlenlehre*), as 'parts or branches of psychology,' though we should have no numbers without counting, no sums without addition, no products without multiplication, etc. (LU 170 [179])

It is implausible to think that Husserl would make this categorical statement while at the same time admitting that he had himself psychologized numbers a few years earlier.

The interpretation offered here is confirmed by an examination of the published volume of PA itself. We see that Husserl does not in fact reduce numbers to mental phenomena in that work. The basic thought of PA is that numbers can be presented both authentically and symbolically (cf. PA 6f., 190ff.). But this would be inconceivable if numbers were regarded as identical with the acts of presentation themselves. Frege himself makes this latter point in his review of PA. Citing Husserl's statement that 'the concepts are given to us, if not in an *authentic* manner, then at least in a *symbolic* one' (PA 192), Frege writes:

> Here the concepts appear as something objective and the difference between authentic and symbolic refers only to their givenness ... The number is not my presentation; rather, I form one or more presentations *of* one and the same number, or try to at least.[90]

There must be a difference between number and the act of counting, between numbers and the presentations that we have of them; otherwise there would be nothing to be given as *the same* in the different acts of presentation in question.[91] If Husserl had not implicitly recognized the 'objectivity' of number, he could not have taken the distinction between symbolic and authentic presentation as the very basis of his early approach to the philosophy of mathematics.[92]

To be sure, Husserl does not say outright in PA that numbers are objects.[93] In this early work he reserves the term object for the real, spatially located and extended objects of the 'external world.' And it is plain that numbers are not objects in this sense; numbers are manifestly 'irreal,' to use Husserl's later term.[94] But if Husserl does not designate numbers as objects in PA, he does at least consider them to be 'concepts.' And concepts, as he makes clear, are not purely psychological entities. A concept may in some sense be the 'content' of a presentation, but it is not identical with the presentation. Different presentations can have, as Husserl says, the same 'logical content' (cf. PA 218). In Chapter II of PA, Husserl gives an illuminating example of the difference between a presentation and its content (PA 31f.). Suppose we have an articulate presentation of a group of objects (A, B, C, D) – that is, a presentation generated by focusing first on A, then on B, C, and D in turn. To describe the presentation itself, as a purely mental phenomenon, we would have to mention the temporal order in which the objects in question were run through. D, the last object to be focused upon, is presented in a different way from A, the first object. If we were now to run through the same group of objects in a different order (say, starting with D and ending with A), the mental phenomenon, the act of presentation itself, would be quite different. But we must distinguish, Husserl says, 'between the [mental] phenomenon as such and that for which it serves us or that which it signifies to us, and accordingly also between the psychological description of a phenomenon and the statement of its *meaning*' (PA 31). The meaning or 'logical content' of the presentations in question remains the same regardless of the order in which the objects are run through. In each case, the content of the presentation is simply the group of objects (A, B, C, D) 'without regard to the temporal differences and the temporal sequence based on it.' In other words, the 'content' of a presentation is not really an immanent part of the presentation itself.

For the Husserl of PA, numbers are not presentations; they are rather the concepts which are 'contained' in certain presentations. He sees and even emphasizes that the same concept can be contained in many different presentations. Of course, Husserl thinks it is important in the philosophy of mathematics to study the presentation of number. But this does not signify that he denies the distinction between contents and presentations, or that he reduces numbers to mental phenomena. As we shall see in detail in the course of this study, there are many reasons to be dissatisfied with the account which Husserl gives in PA of the presentation of number. But one cannot justly accuse him of 'psychologizing' numbers in a pernicious sense. The difficulties are in the execution of the analyses, not primarily in the methodological conception which underlies them.

We are led then to a somewhat unconventional but nonetheless compelling conclusion regarding Husserl's admitted early psychologism in the philosophy of mathematics. His philosophy of mathematics had indeed been genuinely psychologistic, but this fact does not have the implications that it is usually taken to have.

In particular, it does *not* imply that the analyses which make up the bulk of the published volume of PA were psychologistic in the pernicious sense criticized in the Prolegomena to LU. When Husserl writes in the Prolegomena of his own earlier psychologism, he is referring not primarily to the published volume of PA, but rather to the view of logic which had, initially at least, underlain his preparations for the second volume.[95] Paradoxically, it was not the 'psychological' studies of Volume I of PA that he later came to reject as psychologistic,[96] but rather the 'logical' studies he had planned for Volume II. Far from being an example of an absurd and pernicious psychologism, the study presented in the published volume of PA is in fact, as we shall see, a study of the intentional structures at work in our dealings with number. Such a study is not only compatible with Husserl's mature approach to the philosophy of mathematics; it is essential to it.

NOTES

1 For an explanation of the abbreviations used in this study, see the 'Note on Abbreviations,' p. 139.

2 Like other fields of modern mathematics, the discipline or area known as 'analysis' is difficult to define with precision. It is often said to be one of three major areas of mathematics, the other two being algebra and geometry (cf. H. Eves and C.V. Newsom, *An Introduction to the Foundations and Fundamental Concepts of Mathematics* [New York: Holt, Rinehart and Winston, 1965], pp. 196f.). Generally speaking, analysis consists of the calculus of integrals and differentials together with several other branches of mathematics which are in various ways related to the calculus (ibid., p. 197). Husserl's own specialization was in a subfield known as the calculus of variations. His dissertation, entitled 'Beiträge zur Theorie der Variationsrechnung,' raises no philosophical issues and contains little that is of interest today.

3 For a readable discussion of the relation between calculus and the natural sciences, especially mechanics, see M. Kline, *Mathematics in Western Culture* (New York: Oxford University Press, 1964), pp. 214-233.

4 Eves and Newsom, *Foundations and Fundamental Concepts of Mathematics*, p. 198.

5 C.B. Boyer, *A History of Mathematics* (New York: Wiley & Sons, 1968), p. 470; Kline, *Mathematics in Western Culture*, p. 232.

6 Cf. Boyer, *History of Mathematics*, p. 502.

7 Eves and Newsom, *Foundations and Fundamental Concepts of Mathematics*, p. 202.

8 Ibid.

9 For a recent version of such an approach, cf. ibid., pp. 216ff. On Weierstrass' own version, see C.B. Boyer, *The History of the Calculus and its Conceptual Development* (New York: Dover, 1959), pp. 285f.

10 K. Schuhmann, *Husserl-Chronik. Denk- und Lebensweg Edmund Husserls* (The Hague: Nijhoff, 1977), pp. 6-9; L. Eley, 'Einleitung des Herausgebers,' PA xxi-xxv.

11 Weierstrass grew ill the semester after Husserl had arrived, thus prompting the latter to move back to Vienna (cf. PA xxii).

12 Schuhmann, *Husserl-Chronik*, p. 7. Leopold Kronecker, another of Husserl's mathematics teachers in Berlin, is also mentioned here.

13 O. Becker, 'The Philosophy of Edmund Husserl,' tr. by R.O. Elveton, in: *The Phenomenology of Husserl. Selected Critical Readings*, ed. by R.O. Elveton (Chicago: Quadrangle Books, 1970), p. 40. Cf. Schuhmann, *Husserl-Chronik*, p. 345.

14 Schuhmann, *Husserl-Chronik*, p. 7.

15 Ibid.

16 Cf. Eley, 'Einleitung,' PA xxiv n. Boyer writes that 'Weierstrass did not go into the nature of the whole number itself, but began with the concept of whole number as an aggregate of units enjoying one characteristic property in common ...' (*History of the Calculus*, p. 285). But from Husserl's notes, it seems that Weierstrass did make at least a brief attempt to explain whole numbers in terms of counting.
17 Cf. Eley, 'Einleitung,' PA xxiv n.
18 In a letter to Natorp written in 1901, Husserl mentions 'the time from 1886/93' as the period of his most itense work in the philosophy of mathematics; cf. Schuhmann, *Husserl-Chronik*, p. 17.
19 For an account of Husserl's relations with Stumpf, see Herbert Spiegelberg, *The Phenomenological Movement. A Historical Introduction*, 2 vols. (The Hague: Nijhoff, 1971), Volume I, pp. 54, 65f. Third revised edition, 1982.
20 See PA 289-338. Hereafter this essay will be referred to as 'BdZ.'
21 As we shall see below, Husserl's *Philosophie der Arithmetik* (1891) could be considered an expanded version of BdZ; it also contains some essentially new developments, however.
22 Apparently the lecture course was entitled 'Einleitung in die Erkenntnistheorie und Metaphysik'; cf. Schuhmann, *Husserl-Chronik*, p. 23. The manuscript in question (K I 28) must be Husserl's lecture notes for this course, for we find the date '22.XI.87' on p. 24a. This manuscript contains what seem to be preliminary studies for the later work *Philosophie der Arithmetik*. Of particular interest is the material pertaining to Chapters V and X through XIII of the latter work. Much of the material in Chapters X and XII of PA is to be found also in this early manuscript. Chapters V, XI and XIII are represented by a paragraph or at least a few sentences each. There is nothing in the manuscript corresponding to Chapters VI-IX of PA.
23 On Husserl's relation to Weierstrass, see D. Willard, 'Husserl on a Logic That Failed,' paper presented at the Eastern Division meeting of the American Philosophical Association, Washington, D.C., December 1978, pp. 12f.; also B. Picker, 'Die Bedeutung der Mathematik für die Philosophie Edmund Husserls,' *Philosophia naturalis* 7 (1961-1962), pp. 269f., 291, 345.
24 The remark of Husserl's quoted above, p. 3, suggests that he had learned from Weierstrass that analysis was to a great extent 'a mixture of rational thinking and irrational instinct and knack.'
25 Cf. PA 339, Thesis VII: 'Die logische Berechtigung der Anwendung irrationaler und imaginärer Zahlen in allen mathematischen Gebieten ist bisher noch nicht erwiesen.' This thesis is one of several that Husserl defended in a formal disputation in the summer of 1887.
26 Ms K I 28/27a-27b.
27 'Kein Vernünftiger zweifelt an der Richtigkeit ihrer Resultate, kein Naturforscher oder Techniker besinnt sich, dieselben als Werkzeuge der Erforschung und Beherrschung der Natur, wo immer es möglich ist, anzuwenden. Und doch! Wie schlimm steht es mit der Logik der Fundamente dieser Wissenschaft. Es ist unglaublich, und doch unzweifelheft, daß es keinen einzigen unter den Grundbegriffen der Analysis gibt, in betreff dessen Klarheit und Einsicht bestände' (Ms. K I 28/32a).
28 Ms. K I 28/30a.
29 Ibid.
30 Presumably, it was because of this more radical demand that Husserl could not be satisfied solely by Weierstrass' own remarkable achievements in the field of mathematics. In the 1887/88 lecture manuscript, after alluding to these achievements (as well as to those of Cantor and Dedekind), Husserl goes on to say that 'however great the arithmetical progress was ... it did not seem that even they had as yet eliminated all the logical difficulties' (Ms. K I 28/30a).
31 Ms. K I 28/25a.
32 '... diese Beschränkung auf speziellere und speziellere Arbeitsgebiete ist nichts, was den Wert oder die Würde ausmacht. Es ist nur ein notwendiges Übel. Der vollkommene Forscher, der auch ein vollkommener Mensch zu sein strebt, wird darum nie die Beziehung der Wissenschaft zu den allgemeineren und höheren Erkenntniszielen der Menschheit außerAugen lassen

dürfen. Die berusfmässige Beschränkung auf ein enges Fach ist notwendig; aber darin ganz und gar aufgehen ist tadelnswert. Und besonders tadelnswert wird uns derjenige erscheinen müssen, dem selbst die allgemeineren Fragen gleichgültig sind, welche die Fundamenta seiner Wissenschaft, und deren Wert und Würde, sowie deren Stellung im Reiche der menschlichen Erkenntnis überhaupt betreffen' (Ms. K I 28/25a).

33 In the Introduction to BdZ, Husserl mentions metaphysics and logic (PA 289), but goes on to say that the results of psychological studies of our presentation of 'space, time, number, the continuum, etc.' 'must be of significance for metaphysics and logic' (PA 292).

35 On the distinction between genetic and descriptive psychology in Brentano, see T. de Boer, *The Development of Husserl's Thought*, tr. by T. Plantinga (The Hague: Nijhoff, 1978), pp. 52-61; also Spiegelberg, *The Phenomenological Movement*, Volume I, pp. 36-38.

36 De Boer, *Development of Husserl's Thought*, p. 54f., 58f., 77-82.

37 Cf. PA 328, where Husserl refers explicitly to 'descriptive psychology' (*beschreibende Psychologie*); also de Boer, *Development of Husserl's Thought*, pp. 60f., 82-90.

38 'Die Mathematik hat es immer und überall nur zu tun mit rein logischen Abhängigkeiten der Größen und Lagenverhältnisse, die Naturwissenschaften mit den realen Verhältnissen der Dinge zueinander (erstere unmittelbar gegeben, letztere hypothetisch und erschlossen). Jedes mathematische Gesetz unterliegt darum dem Satz vom Widerspruch, in der Art, daß wer es leugnete, eine Absurdität behauptete. Jedes Naturgesetz unterliegt dieser Art nicht dem Satz des Widerspruchs, sondern nur den Lehrsätzen der Induktion. Es ist ein induktives Gesetz, und wer es leugnet verfällt nur in eine höchst unwahrscheinliche, bestenfalls unendlich unwahrscheinliche Behauptung' (Ms. K I 28/16a).

39 For a full discussion of the issue of Husserl's early psychologism, see pp. 19-23.

40 Cf. Eley, 'Einleitung,' PA xxiv.

41 PA 193 n. In light of this explicit statement of indebtedness, B. Picker's claim that Husserl's interest in symbolic presentation was 'stimulated by Gauss' intuitive representation (*Veranschaulichung*) of the imaginary numbers' (Picker, 'Bedeutung der Mathematik,' p. 292) seems highly implausible.

42 Ms. K I 28/8a-12b. Cf. above, p. 24, n. 22.

43 Ms. K I 28/10a-12b.

44 See below, pp. 38f.

45 PA 190; cf. Ms. K I 28/9b-10a.

46 The forthcoming edition of Husserl's *Studia Mathematica* in the *Husserliana* series may contain materials which help to resolve this issue.

47 The justification for this latter date is the letter cited by W. Biemel in his article 'The Decisive Phases in the Development of Husserl's Philosophy,' tr. by R.O. Elveton, in *The Phenomenology of Husserl*, ed. Elveton, p. 154. A manuscript bearing the date January, 1889, however, already seems to express the new point of view. It was Biemel who first called attention to this development and its significance, though without studying it in detail. More recently, D. Willard has argued that a proper understanding of this development is essential to a full comprehension of the early Husserl's thought; see 'Husserl on a Logic That Failed,' pp. 21-30.

48 PA 1-283. The second volume never appeared in print.

49 Willard however has called attention to it; see his 'Husserl on a Logic That Failed,' pp. 23, 25-28.

50 M. Farber, for example, describes PA as 'a well-organized and, in the main, remarkably clear treatise'; *The Foundation of Phenomenology* (Albany: S.U.N.Y. Press, 1968), p. 25.

51 According to this statement, the printed version of BdZ does not represent the whole of Husserl's habilitation essay, but only a part. This means that much more than the first few chapters of PA (which correspond to the *printed* portion of the habilitation essay, i.e. to BdZ) could also have been taken 'almost verbatim' from a work written in 1887. However, since the original habilitation essay has not been found, it is impossible to assess the extent to which this is true.

52 E. Husserl, 'Persönliche Aufzeichnungen,' ed. by W. Biemel, *Philosophy and Phenomenological Research* XVI (1956), p. 294. The quoted passage is in substantial agreement with a remark made by Husserl in a letter written at the time of PA's publication: 'Das Werk [PA] ist zum größten Theile eine Arbeit früherer Jahre und ist unter so vielfachen Widrigkeiten und Hemmungen entstanden, daß es schwerlich den Eindruck eines harmonisch ausgeglichenen Ganzen machen wird'; letter to Meinong of 22 May 1891 in: *Philosophenbriefe. Aus der wissenschaftlichen Korrespondenz von Alexius Meinong*, ed. by R. Kindinger (Graz, Austria: Akademische Druck- und Verlagsanstalt, 1965), p. 100.
53 PA 294f.; cf. above, p. 5.
54 Cf. Willard, 'Husserl on a Logic That Failed,' p. 27, n. 56.
55 Although much of the passage under consideration here is modelled on a similar passage in BdZ (PA 296f.), the allusion to the possible 'untenability' of the Weierstrassian view is found only in PA itself.
56 Willard speculates that Husserl may have sent some parts of PA to the printer before those parts in which he expresses his subsequent views were completed; 'Husserl on a Logic That Failed,' p. 28, n. 57. But since PA was not published until the spring of 1891, whereas Husserl had already expressed the new view as early as February, 1890, this seems improbable.
57 Letter of 13 February 1890, cited in Biemel, 'The Decisive Phases,' p. 154.
58 Ibid.
59 Ibid. Husserl emphasizes the term 'formal' in this sentence apparently to stress that he has formal or symbolic logic in mind, rather than philosophical logic or epistemology. Husserl used the term logic in several different senses at this early stage of his career, as well as later on.
60 Cf. de Boer, *Development of Husserl's Thought*, pp. 91-93.
61 F. Brentano, *Die Lehre vom richtigen Urteil*, ed. by F. Mayer-Hillebrand (Bern: Francke Verlag, 1956), pp. 1f., 3, 5. This posthumously published work is based on a lecture course which Brentano gave in several different semesters, including the winter semester of 1884/85, when Husserl was his student.
62 Cf. de Boer, *Development of Husserl's Thought*, pp. 92f.
63 Husserl, letter of 13 February 1890, cited in Biemel, 'The Decisive Phases,' p. 154. The complete text of this letter has not been published; it is possible that Husserl's discussion of 'formal logic' was somewhat more extensive than we have indicated.
64 PA 340-373. On the dating of this manuscript, see the editor's 'Textkritische Anmerkungen,' PA 524. R. Verdenal studies some aspects of this diffuse but important manuscript in an article entitled 'La semiotique de Husserl: la logique des signes (A propos de certains inédits),' *Les Etudes Philosophiques* 4 (1973), pp. 553-564; further study would seem warranted however.
65 PA 373; emphasis added.
66 Cf. Schuhmann, *Husserl-Chronik*, p. 43.
67 Biemel points out in 'The Decisive Phases' (p. 155) that Husserl's transition to logic was 'not at all arbitrary but the result of a coherent development.' For his early concern with the nature of analysis had led him to the view that it was 'a piece of formal logic,' as we have seen; this view in turn suggested a need for yet more penetrating investigations of the nature of logic.
68 The edition cited in the present study is that published in 1968 by Max Niemeyer Verlag, Tübingen. This edition (hereafter cited as 'LU') is the fifth edition of Volumes I and II/1, the fourth of Volume II/2. The English translation as *Logical Investigations* by J.N. Findlay, two volumes (London: Routledge and Kegan Paul, 1970), will be cited in brackets or parentheses following citations of the German edition.
69 The German edition of *Ideen* cited herein is that published as Volume III of the *Husserliana* series, ed. by W. Biemel (The Hague: Nijhoff, 1950; new edition by Karl Schuhmann, 1976). The translation by W.R. Boyce-Gibson (New York: Collier Books, 1972) will be cited in brackets or parentheses following citations of the German edition.
70 The German edition cited herein as 'FTL' is published as Volume XVII of the *Husserliana* series, ed. by P. Janssen (The Hague: Nijhoff, 1974). The English translation as *Formal and*

Transcendental Logic by D. Cairns (The Hague: Nijhoff, 1969) will be cited in brackets or parentheses following citations of the German edition.
71 The first German edition was published as Volume VI of the *Husserliana* series, ed. by W. Biemel (The Hague: Nijhoff, 1954). The English translation as *The Crisis of European Sciences and Transcendental Phenomenology* by D. Carr (Evanston, Ill.: Northwestern University Press, 1970) will be cited in parentheses following citations of the German edition.
72 See PA 430-451. According to the editor, the manuscript published in PA may not be the complete text of the lecture Husserl actually delivered; 'Textkritische Anmerkungen,' PA 549f. For more information on this lecture, see *Ideen* 168 n. (188 n.); FTL 101 (97); and Schuhmann, *Husserl-Chronik*, p. 68 n.
73 Ms F I 16. The course was entitled 'Einführung in die Logik.'
74 LU II/1 249 (241). Husserl quotes this passage verbatim in FTL with the comment that he 'cannot improve on it' (FTL 94f. [91]).
75 LU I 250, 252 (242, 243); FTL 81, 103 (77, 99).
76 In a footnote in *Ideen*, Husserl remarks that he had made use of the concept of manifolds 'already in the early 1890's' (*schon zu Anfang der 90er Jahre) (Ideen* 168 n. [188 n.]). Manuscript evidence seems to confirm this; cf. Schuhmann, *Husserl-Chronik*, p. 31, where manuscripts dating from 1891/1892 on 'Mannigfaltigkeitslehre, Reihen, zyklische Reihen' are mentioned. It schould be noted however that, according to the remark in *Ideen*, it was only 'in the *Studies in the Theory of the Formal Mathematical Disciplines* which [Husserl] had planned as the sequel to [his] *Philosophy of Arithmetic*' that he had made use of the concept of manifolds. The term 'manifold' is not used in the technical Husserlian sense in the published volume of PA. Husserl does say in this volume that numbers constitute 'a manifold of determinate concepts that are most sharply delineated from one another' and that 'the region of number embraces an *unlimited manifold of species*' (PA 81, 222 f.). But he is here discussing the 'original' number region, not a region which is defined only as the objective correlate of a theory form. Farber claims to find an 'adumbration' of Husserl's later notion of manifolds in the discussion of 'infinite aggregates' at PA 218 ff.; *Foundation of Phenomenology*, pp. 50, 59. In this however he seems to have been misled by the non-technical uses of the term 'manifold' cited above, and perhaps also by Boyce-Gibson's mistranslation of the footnote on *Ideen* 168 (188) (Boyce-Gibson has 'towards the beginning of 1890' instead of 'in the early 1890's'; cf. Farber, *Foundation of Phenomenology*, p. 50 n.). As we have pointed out above, a 'manifold' in Husserl's later technical sense is something very different from a set or aggregate. The only genuine adumbration of this notion in PA is to be found in the discussion of 0 and 1 as numbers at PA 131 ff. It is interesting in this connection to note Husserl's brief reference in the Preface to Gauss' essay on biquadratic remainders (PA 8). Bachelard points out that Riemann, whose use of the term 'manifold' seems to have inspired Husserl's own technical use of this term, expressed a certain indebtedness to this same essay by Gauss; Bachelard, *Study of Husserl's 'Logic'*, p. 57 n.
77 Husserl explains that he understands the term 'imaginary' in a very broad sense, so as to include 'also the negative, indeed even fractions, irrational numbers, etc.' (PA 432f.).
78 In a letter written around this time Husserl stated that at least one of the lectures was taken 'from old mathematical philosophical manuscripts'; Schuhmann, *Husserl-Chronik*, p. 68.
79 In the context of the quoted sentence, Husserl alludes to the Preface to PA; cf. FTL 103 n. (99 n.).
80 See the subtitles of the two works (PA 1, 289); also PA 6f., 295.
81 PA 6f., 77ff., 298, 334ff.
82 G. Frege, 'Rezension von: E.G. Husserl, Philosophie der Arithmetik, I,' in: *Kleine Schriften*, ed. by I. Angelelli (Darmstadt: Wissenschaftliche Buchgesellschaft, 1967), p. 192.
83 E. Pivcevic, for example, claims that Husserl's early position regarding the concept of number and the need to clarify it with the help of psychology was 'subsequently criticized by Husserl himself as 'psychologistic''; *Husserl and Phenomenology* (London: Hutchinson & Co., 1970), p. 24. See also Farber, *Foundation of Phenomenology*, pp. 18, 21; Biemel, 'The Decisive Phases,' p. 152; D. Føllesdal, 'Husserl und Frege. Ein Beitrag zur Beleuchtung der Entstehung

der phänomenologischen Philosophie,: *Avhandlinger utgitt av Det Norske Videnskaps-Akademi i Oslo. Hist.-Filos. Klasse* II (1958), No. 2, pp. 48ff.; F. Bosio, 'Gli inizi della fenomenologia. La *Filosofia dell'Aritmetica,' Aut Aut* 70 (July 1962), p. 295; R. Sokolowski, *The Formation of Husserl's Concept of Constitution* (The Hague: Nijhoff, 1970), pp. 6, 30; J. Huertes-Jourda, 'On the Threshold of Phenomenology. A Study of Edmund Husserl's *Philosophie der Arithmetik*,' typewritten.

84 In LU, Husserl writes that he still finds his earlier psychological studies of the 'origin of mathematical presentations' to be 'clear and instructive' (*klar und lehrreich*) (LU I vii [42]). In FTL, he goes so far as to say that PA was in fact a 'phenomenologico-constitutive investigation' (FTL 91 [87]); see below, p. 37 ff.

85 It has sometimes been acknowledged, even by those who consider the study of number in the published volume of PA to be 'psychologistic' in some sense, that the psychologism in question here is not the pernicious, thoroughly vitiating kind criticized by Husserl in LU. See Becker, 'The Philosophy of Edmund Husserl,' p. 41; Sokolowski, *The Formation of Husserl's Concept of Constitution*, pp. 15ff., 35; R. Torretti, 'La Filosofia de la Aritmetica de Husserl,' *Studi Internazionali di Filosofia* 4 (1972), p. 190. Others have argued that PA at least contains the germs of an 'overcoming' of psychologism. See L. Landgrebe, 'Husserls Phänomenologie. Das Problem einer ursprünglichen Erfahrung' (Gutersloh, West Germany: Verlagshaus Gerd Mohn, 1967), p. 13; Biemel, 'The Decisive Phases,' p. 152; Bosio, 'Gli inizi della fenomenologia,' pp. 300, 305ff.; J. Huertes-Jourda, 'On the Threshold of Phenomenology.' But what is missing in these studies is a satisfactory account of just what Husserl *did* mean when he spoke unambiguously in LU of his own earlier 'psychologistic view of the philosophy of mathematics.'

86 See especially LU I 47ff. (86ff.).

87 In an 1897 letter, Husserl described an early draft of the Prolegomena as an essay 'which is directed against the subjectivistic-psychologistic logic of our time (and thus against the standpoint which I earlier advocated as a student of Brentano's).' Schuhmann, *Husserl-Chronik*, p. 49. This remark should not be misinterpreted to mean that Husserl was at one time guilty of the sort of psychologism which he later criticized in Mill, Sigwart and others. The latter theorists sought to incorporate logic into 'genetic' or empirical psychology; Husserl, like his teacher Brentano, had regarded it as founded merely on 'descriptive' psychology.

88 The psychologism criticized in LU should not be confused with what Husserl later referred to as 'transcendental psychologism' (FTL 257ff. [250ff.]). Husserl's early works are undoubtedly to be reproached for psychologism of this latter sort; for that matter, even LU does not escape it, as Husserl himself noted in FTL (FTL 160f. [152]). But 'transcendental psychologism' is not pernicious or absurd in the way the psychologism criticized in LU is. It consists simply in a lack of clarity about the being of consciousness, a failure to appropriate and revaluate the results of descriptive psychology from a transcendental point of view; see FTL 261f. (254f.).

89 This interpretation is consistent with Husserl's remarks in the Foreword to LU. Here we find the author expressing continued satisfaction with his earlier psychological analysis of the 'origin of mathematical presentations' (LU I vii [42]). Theoretical difficulties had arisen for him, he reports, when he moved from the study of such presentations to logical problems strictly speaking. It was only when he began to investigate 'the logical unity of the thought-content (the unity of theory)' of mathematics that he became dissatisfied with his earlier point of view. Clearly, such problems had not been of central concern in the published volume of PA; cf. Torretti, 'La Filosofia de la Aritmetica de Husserl,' p. 185.

90 Frege, 'Rezension,' pp. 191f. Frege however does not see that the distinction between authentic and symbolic modes of givenness is central to PA as a whole. He is thus forced to interpret the work as basically inconsistent. He holds that the 'fog' of psychologism darkens the early sections, but that the 'sun of truth' breaks through in Part Two when Husserl begins to discuss the difference between the symbolic and authentic presentation of number. This is a

misinterpretation of PA, though it is a natural one in view of the fact that Husserl does not make the authentic-symbolic distinction thematic at an early point in the work. The misinterpretation is particularly unfortunate, for it enables Frege to avoid confronting the real issues separating his approach from Husserl's.

91 Cf. Landgrebe, 'Husserls Phänomenologie und die Motive zu ihrer Umbildung,' pp. 11ff.

92 Cf. LU II/1 282 n. (480 n.); FTL 90f. (86f.). Of course, to say that Husserl implicitly recognized the objectivity of number is not to say that he could account for it philosophically. This seems to be the sense of Husserl's comment in the Foreword to LU that he had been unable to clarify the relationship 'between the subjectivity of knowing and the objectivity of the content known' (LU I vii [42]).

93 But cf. PA 295, where Husserl compares number to space, 'the object of geometers.'

94 H. Pietersma cites PA 317 as a passage that 'leaves no doubt' about the fact that 'Husserl ultimately identified numbers with a mental power'; 'Husserl and Frege,' *Archiv für Geschichte der Philosophie* 49 (1967), p. 316. But Husserl says in this passage that numbers are the 'results' of certain mental activities, not that they are identical with the mental activities themselves. His point is that numbers cannot be 'found ready-made' (*fertig vorgefunden*), as the objects of sense perception can be. Numbers may be as objective as the North Sea, but they are not as 'real.'

95 Some allusions to this view of logic are perhaps to be observed in the Preface to PA and in the final chapter of the published volume – a chapter which examines the 'logical sources of arithmetic' (PA 256ff.). But it was only in the planned second volume of the work that logical questions were to have become a central theme in their own right.

96 The fact that these analyses are labeled 'psychological' does not, of course, prove that they commit the error criticized as psychologism in LU. For even the Husserl of LU's first edition demands a 'descriptive psychological' analysis of logical and mathematical concepts; see Findlay's translation of LU, p. 262f. It was only after the publication of LU that Husserl realized that what he had been doing was not really 'psychology' at all. He had not been studying the human mind, but rather the being of the things themselves.

CHAPTER II

HUSSERL AND THE CONCEPT OF NUMBER

1. THE DEFINITION OF NUMBER

What is number? What do we really mean by 'number' in the most fundamental sense of the word? In Chapters II through IV of this study, we shall be concerned with Husserl's attempt to answer these ancient and difficult questions with the help of phenomenology. We shall see how Husserl uses phenomenological analyses to explain what numbers themselves truly are. In Chapters III and IV, we shall examine his analyses in considerable detail. In the present chapter, however, we shall deal with some preliminary questions relatings to Husserl's methodology. Here we shall seek only to explain in general terms what is meant by a phenomenological study of number.

Number has been defined in many ways. One of the most ancient and influential definitions in the history of Western thought is that which was given by Euclid at the beginning of Book VII of the *Elements*: according to Euclid, a number (*arithmos*) is a 'multitude consisting of units' (*to ek monadon synkeimenon plethos*).[1] Now it could be said that this definition was accepted as essentially correct by most theorists even into the modern period. Even in the late nineteenth century, the great mathematician Weierstrass could repeat it at the beginning of his lectures on analytic functions.[2] At about the same time, however, the classical Euclidean definition was being challenged in a radical way by the German logician Gottlob Frege. In his influential work *Die Grundlagen der Arithmetik*, Frege proposed a new definition of number in terms of one-to-one correspondence or 'equality' (*Gleichzahligkeit*). According to Frege, 'the number which belongs to the concept *F* is the extension of the concept 'equal to the concept *F*."[3] This Fregean definition of number has subsequently come to be accepted by many mathematicians and philosophers, just as the Euclidean definition had been in earlier centuries. Let us therefore begin our discussion of Husserl's approach to the concept of number by examining his relation to these two influential definitions.

Husserl devotes careful attention in his early works on the philosophy of arithmetic to both of the definitions in question. At the same time, however, he denies that the concept of number can be clarified adequately through any definition by itself. Indeed, he says at one point in PA that the concept of number is 'barely'

See notes at the end of this chapter.

definable at all.[4] In Husserl's view, the crucial issue is not just how we are to define number, but rather how we are to take the definition that we give. The primary thing is not the words we use in the definition, but rather the understanding that we have of those words. As Husserl himself puts it, 'the difficulty lies in the phenomena, their correct description, analysis and interpretation; only with regard to them can we achieve insight into the essence of the number concepts' (PA 129). It is through such 'description, analysis and interpretation' of the 'phenomena' with which number is involved that Husserl seeks to clarify the concept of number. His goal is not just to define number, but to show how number makes its 'appearance' in the world in which we live. In the most general terms, this is what is meant by a 'phenomenological' study of the number concept: a descriptive study of the manner in which number makes its appearance in the world in which we live.

But even if the ultimate clarification of the number concept requires phenomenological analysis, this does not by any means imply that the definition of number is a matter of indifference. Even a phenomenological study of number must be guided by some initial understanding of what number is. Without some such understanding, we could not know which phenomena were in need of description and analysis. A suitable definition can guide us, so to speak, to the phenomena. But how then is number to be defined? Should we accept the Euclidean definition of number as a 'multitude of units'? Or should we follow Frege and define number as that which is the same when two concepts are equal?

Husserl rejects both of these definitions. He seems to find something of significance in each one, but he himself defines number in a different way. According to Husserl, numbers in the fundamental sense are to be defined as *determinate multitudes (bestimmte Vielheiten)*.[5] It is this understanding of number which guides him in his attempt to clarify the number concept. When he studies number with the help of phenomenology, he seeks to show just how it is that 'determinate multitudes' make their appearance in the world in which we live.

In defining number as determinate multitude, Husserl stands in partial agreement and partial disagreement with both Euclid and Frege. An understanding of the respective agreements and disagreements will help us to appreciate the sense of the Husserlian definition of number.

Let us consider the Euclidean definition first. Husserl agrees with this definition in regard to what it cites as the 'genus' of number; he agrees that numbers are in some sense 'multitudes.' 'Where there is talk of a determinate number, one can always speak of a multitude ...' (PA 14). 'The concept of number embraces ... the same concrete phenomena as the concept of multitude' (PA 15). But Husserl disagrees with the Euclidean definition in regard to what it cites, or seems to cite, as the 'specific difference' of number. According to Husserl, 'it is not a feature somehow characteristic of that multitude which we call number that it consists of units' (PA 128). In fact, Husserl contends, *all* genuine multitudes are multitudes of units or 'ones.' 'Multitude' and 'unit' or 'one' are simply correlative terms (PA 84, 128). What is it that makes up the specific difference of number, then, if it is not the presence of units? Multitude as such has a 'vague indeterminateness' about it, but when we speak of number as distinct from multitude,[6] we have in mind a concept which includes two, three, four, etc., as its specifications (PA 14f., 83). What distinguishes number from multitude in general is therefore the presence of 'a precisely determinate *how many*' (PA 83); numbers are not just multitudes, but 'modes of how-many' (FTL 81 [77]). The concept of number implies that what is 'many' has been determined more precisely. This is what Husserl seeks to bring out through his correction of the classical Euclidean definition.

It might be said that the considerations which lead Husserl to make 'determinateness' the specific difference of number are similar in nature to those which lead Frege to define number as he does. Husserl's definition stresses the relational character of the number concept, the fact that number involves the determination of multitudes in relation to each other. Something similar is emphasized in Frege's definition of number as that which is the same when two concepts are equal. But Frege himself of course denies that 'multitude' is the genus of number. Frege contends that an assertion of number is an 'assertion about a concept,' not an assertion about a multitude.[7] Frege in effect severs the essential connection between number and multitude that had been asserted in the Euclidean definition. As Husserl points out, Frege's rejection of the latter definition is much more radical than his own (PA 129). Where Husserl rejects only the specific difference asserted in the Euclidean definition, Frege rejects the genus itself.[8]

Two considerations seem especially important in leading Frege to deny that the genus of number is multitude. On the one hand, Frege points out that this definition is applicable only to the numbers in the number series beginning with two. 'All these views suffer from the drawback that the concept will not cover the numbers 0 and 1.'[9] On the other hand, there is according to Frege a certain vagueness and variability in the notion of a multitude. The same multitude or aggregate can be assigned different numbers, depending on how we take it: we can speak either of 'a copse' or of 'five trees,' either of 'four companies' or of 'five hundred men.'[10]

How does Husserl reply to these objections? Although the details of Husserl's treatment of these issues will emerge only later in the course of this study, it will be helpful for us to indicate their general direction even at this early stage. To the first point mentioned above, Husserl replies that zero and one are indeed excluded by this definition, but that their exclusion is entirely appropriate (PA 130ff.). He does not of course deny that zero and one are in *some sense* numbers; for that matter, he does not deny that -1 and $\sqrt{2}$ are in some sense numbers. Husserl's point is only that zero and one are not numbers in the most fundamental sense; they are not numbers in the sense of *Anzahlen.*[11] To clarify the other, derivative senses of number, we must first show what is meant by 'number' in this most basic sense of the word. Without reference to multitudes, this cannot be done.

Husserl's response to Frege's other objection is that we must distinguish between clusters, heaps, etc., and multitudes properly so-called. To say that numbers are multitudes is not to say that they are clusters or heaps. It is only in a multitude or 'aggregate' (*Inbegriff*) that each object is taken as 'one.' 'But the objects by themselves do not determine the aggregate. The same objects can be presented in different aggregate forms' (PA 163f.). And thus, despite the variability noted by Frege, 'the number is univocally determined if the aggregate is determined' (PA 163). It is of course necessary to be precise about the difference between a cluster or heap, on the one hand, and a multitude or aggregate, on the other. But Husserl contends that this distinction can be sustained.

In the final analysis, the chief difference between Frege and Husserl is that Husserl but not Frege seeks a definition which would explicate 'what we really mean by number.' Frege contends that the 'mathematician' is concerned not with the sense or content of a concept such as number, but only with 'the thing itself, the referent.'[12] For the mathematician, he admits, any definition is plausible if it correctly delimits the 'extension' of the concept being defined.[13] Systematic considerations are necessary in order to determine which of several plausible definitions is to be selected; a definition shows its worth, Frege says, by proving useful.[14] What Husserl expects of a definition is quite different. In his view, definition has a philosophical function as well as a mathematical one in the narrower sense. A definition of number should, on Husserl's view, bring to light the sense or content of this concept (PA 117, 124). It should, in other words, help us to understand what we really mean by number. Despite Frege's claim to the contrary, this expectation is in no way undermined by a sharp distinction between number and the presentation of number.[15]

2. THE ORIGIN OF NUMBER AS A PHENOMENOLOGICAL PROBLEM

Husserl defines 'number as 'determinate multitude,' but he also insists on the need to go beyond any mere definition of number and study the 'phenomena' with which number is involved. He seeks to study the manner in which number makes its appearance in the world in which we live. We must now attempt to characterize Husserl's 'phenomenological' approach to the problem of number more carefully.

Husserl sometimes describes his approach as one in which the 'origin' of number is studied. In LU, for example, he states that to obtain 'insight into the essence' of a concept such as number, we must come to understand its 'phenomenological origin.'[16] Now the term 'origin' is susceptible to a variety of possible misunderstandings. A study of the origin of a concept might be taken to mean a study of the *historical* sources from which this concept derives. But it is clear that Husserl is not using the term 'origin' in this sense. As he remarks in his late essay entitled 'The Origin of Geometry,' the question of origins is not a 'philological-historical question' (K 366 [354]). Or the study of the origin of a concept might be taken to mean an inquiry into the *causes* which lead men to form that concept. But this is not Husserl's meaning either. He states in LU, with obvious reference to 'genetic' or explanatory psychology, that 'psychological questions as to the origin of the conceptual presentations' are not of the 'slightest interest' in the present context (LU I 244 [237 f.]).

What then *does* Husserl mean by 'origin'? What is he seeking when he investigates the 'phenomenological origin' of a concept such as number?

It is important to distinguish between two senses of the term 'origin' even within the context of Husserl's own philosophy.[17] Like the well-known empiricists whose works he studied closely, Husserl sometimes uses this term to designate the process of 'abstraction' by which we move from concrete, individual objects to general concepts. In this sense, a study of the origin of a concept might be said to involve two phases: first, a description of a certain class of concrete objects; and second, a description of the process by which the general concept is actually abstracted from the concrete objects. It is primarily in this sense that Husserl uses the term 'origin' in his early writings on the philosophy of arithmetic.[18] In these works he first describes the 'concrete phenomena' that form the basis for our abstraction of the number concept, and then goes on to discuss the process of abstraction itself.[19]

But this empiricist notion of origins is not identical with the specifically phenomenological notion which comes into prominence in Husserl's later writings. In these works, the term 'origin' designates not merely the process by which we move from concrete objects to general concepts, but rather the intentional syntheses by which the concrete objects themselves are first 'constituted' as well.[20] This later sense of origin – origin in the sense of constitution rather than abstraction – does not of course exclude the earlier sense. But it does call attention to a deeper level of the problem of origins. It is this deeper, genuinely phenomenological sense of 'origin' that we have in mind qhen we refer in the present study to the problem of the origin of number. Let us therefore attempt to explain this problem in more detail.

The phenomenological concept of origin is best understood in light of the analyses of evidence, truth and being which Husserl presents in the Sixth Investigation of LU. The background of these analyses – whose richness and complexity can only be suggested here – is the fundamental distinction Husserl draws between 'empty' and 'filled' intentions. Intentional acts are always directed at objects, but they are not always of the same character, as Husserl points out. An act can be an 'empty' one in which the object is indeed meant in any of a variety of ways, but not actually given intuitively. In this case, we might say, the object is intended in its absence.[21] On the other hand, the object can also be given directly and intuitively. Here the act that intends it is not empty, but rather intuitively saturated or 'filled.' The object is intended in its very presence. Now the terms evidence, truth and being must be elucidated, Husserl contends, with reference to the interplay of such empty and filled intentions. Evidence (*Evidenz*) is not just a feeling of assurance, but rather a synthetic act in which an empty intention comes to be intuitively filled. 'The object is not merely meant, but in the strictest sense *given*, and given as it is meant, and made one with our meaning-reference' (LU II/2 122 [765]). Evidence arises, in other words, when we have as present an object that had been or could be intended in its absence.

Now according to Husserl, the concepts of truth and being are both founded on the experience of evidence. They both involve our awareness of the *identity* which underlies the act of evidence. For it is essential to evidence that the object given intuitively be the very same as the object that was meant, and it is possible for us to make not only the given object, but also its 'sameness' a theme of awareness in its own right. Thus Husserl describes what he calls 'being in the sense of truth' as 'an *identity: the full agreement* of what is meant with what is given as such.'[22] To speak of an object's 'being' in this sense is to make reference then to a state of affairs in which the object is grasped as something that could be intended either emptily or in an intuitively filled manner. The word 'being' designates identity in presence and absence.

In a careful study of the central themes of Husserl's philosophy, Ernst Tugendhat has argued persuasively that this analysis of the concepts of evidence, truth and being is the cornerstone upon which Husserl's concept of constitution is built.[23] In studying the 'constitution' of an object, we study its 'being' in the sense alluded to above; that is, we examine the complex of intentional acts within which the object comes to be given with evidence or 'itself-given.' We view the object as something which is intended as 'the same' in an array of empty and filled acts. We study the object as an identity in presence and absence.

It is in this sense that we must understand the specifically phenomenological concept of origin. To study the origin of an object through phenomenology is to bring out the intentional acts – the empty as well as the filled – within which the object emerges as something identically the same. It is to describe the essential features of the acts through which the object is intended as an identity. Now as Husserl points out at length in LU, different kinds of objects lead us to different kinds of acts. The intentional acts required to make a fact or collection intuitively present, for instance, are quite different from those which suffice to make sensuous objects present (LU II/2 128 ff. [773 ff.]). Facts and collections are examples of what Husserl calls 'categorial objects.' To have such objects intuitively present, we must perform a corresponding categorial act – an articulated act of judging or collecting, as the case may be.[24] To study such acts in their difference from and relation to the empty, unfulfilled acts directed at the same categorial objects is to give a phenomenological analysis of the constitutive origins of those objects.

Now numbers, according to Husserl, are also categorial objects.[25] They too are given intuitively only through the performance of certain corresponding categorial activities. When we study the 'origin of number' from a phenomenological point of view, it is these activities, in their difference from and relation to other intentional acts directed at numbers, that we seek to bring to light. In making such a study we seek to exhibit the specific set of presences and absences within which numbers emerge as identities.

What sort of categorial activity is it in which numbers themselves are made intuitively, authentically present? In some of his later works, Husserl calls this activity 'counting' (*Zählen*).[26] Now this is at first somewhat puzzling. When we count the items in a group in the ordinary manner, we run through the items one-by-one, co-ordinating a number word with each item as we do so. But to do this, it seems, we must *already* have some familiarity with numbers. At the very least, we must have learned the number words in their proper sequence. We must also know how to use them to find the number of the group whose items we are counting. But in this case, how can it be said that counting is an activity in which numbers are made present in an 'original' manner? Must there not be a yet more fundamental activity which Husserl is overlooking?

This apparent paradox can be resolved only if we examine Husserl's early studies of the concept of number carefully.[27] In these early studies we find that he in fact uses the term 'counting' in two distinct senses. The everyday counting we have just described is, on Husserl's view, only one type of counting. He describes it as a relatively 'blind,' 'mechanical' and 'sensuously-symbolic' form of counting in which we do not become conscious of the 'conceptual content' of the number words we use (PA 177, 239). In contrast to this everyday counting, the early Husserl also speaks occasionally of what we might call 'authentic counting.'[28] In authentic counting, the 'conceptual content' of the number words we use is in fact activated. The number words are not merely used in a blind, mechanical manner as 'labels' for the items of a group. Rather, they are taken as genuine names whose meanings are thought through in active, articulate fashion. It is only in this authentic counting that a truly original presence of number is achieved.[29]

Since this is clearly the view which Husserl takes in his early works on the philosophy of arithmetic, it is plausible to suppose that he meant something similar in his later works when he spoke of 'counting' as the origin of number. When the later Husserl states that numbers are constituted originally in the act of counting, we suggest, he has in mind not the familiar counting procedure of everyday life, but rather something akin to what he had earlier referred to as 'authentic' counting. But at this point we are led into a controversy concerning the proper interpretation of Husserl's early works. Are they in fact to be taken as studies of the constitutive origin of number at all?

3. THE ORIGIN OF NUMBER IN HUSSERL'S EARLY WRITINGS

In FTL Husserl describes his early work PA as an analysis of the constitutive origins of certain categorical objects, including number:

> In spite of its immaturity as a first book, [PA] presented an initial attempt to go back to the spontaneous activities of collecting and counting, in which collections ('sums,' 'sets') and cardinal numbers are given in the manner of something that is being generated originaliter, and thereby to gain clarity respecting the proper, the authentic, sense of the concepts fundamental to the theory of sets and the theory of cardinal numbers. It was therefore, in my later terminology, a phenomenologico-constitutive investigation; and at the same time it was the first investigation that sought to make 'categorial objectivities' of the first and of higher levels ... understandable on the basis of the 'constituting' intentional activities, as whose productions they make their appearance originaliter, accordingly with full originality of their sense (FTL 90 f. [86 f.]).

Although this retrospective self-interpretation has been accepted by some scholars as basically correct,[30] it has also been challenged as anachronistic and seriously misleading. Theodore de Boer, in a careful and thorough study of Husserl's development, has argued that PA is *not* in fact an analysis of the constitution of number. He contends that PA is an 'analysis of origins' only in the sense that it studies the process by which we abstract certain concepts from the concrete phenomena on which they are based.[31] According to de Boer, the specifically phenomenological concept of origin (origin in the sense of constitution rather than abstraction) is not a factor in this early work.[32] On his interpretation, the 'higher order acts' with reference to which concrete collections and numbers are explained in PA have no 'objective correlates' at all; the acts are themselves the collections and numbers in question, not the constitutive sources of the latter.[33]

This issue is an important one which we must examine carefully. In doing so it will be helpful for us to distinguish two questions: (1) Is the problem of constitutive origins addressed *at all* in PA? (2) Is the problem of the constitutive origin of *number in particular* addressed in PA?

(1) It seems reasonable, we suggest, to answer the first question in the affirmative. However clumsy and unsatisfactory Husserl's treatment of the issue may be, he is in fact trying to bring to light the origins of certain higher-order objects in the specifically phenomenological sense of 'origin' elucidated in the preceding section.[34] It is true, as de Boer points out, that the term 'origin' (*Ursprung* or *Entstehung*) is used in Husserl's early works primarily to designate the process of abstraction. But even this statement must be qualified, for Husserl does in fact use the verb '*entstehen*' at a privotal point in these works with reference to a concrete object rather than a concept. A concrete aggregate 'comes to be' (*entsteht*), he says, when we perform a certain higher-order act (PA 74, 333).

More important however is the fact that an aggregate is understood even by the early Husserl as something which is 'the same' in many different acts. In Chapter I of this study we pointed out that Husserl explicitly distinguishes in PA between the aggregate itself and the acts in which it is presented.[35] A concrete aggregate is not identical with the presenting act; it is rather the latter's 'meaning' (*Bedeutung*) or 'logical content' (PA 31, 309). Moreover, the complex of acts within which the aggregate emerges include those which are 'empty' as well as those which are 'filled.' To be sure, the terms 'empty' and 'filled' are not actually used in PA in the relevant senses. Husserl does however distinguish in this work between 'symbolic' and 'authentic' presentation. Indeed, the very structure of the work is based on this distinction.[36] Husserl explains the distinction as follows:

> A *symbolic* or inauthentic presentation is, as the name itself suggests, a presentation through signs. If a content is not directly given to us as what it is, but only indirectly *through signs that characterize it univocally*, then we have a symbolic rather than an authentic presentation of it (PA 193).

As an example of this distinction, Husserl refers to two different presentations we might have of the outside of a house. We have an authentic presentation if we actually look at the house, a symbolic presentation if someone merely describes its location to us (PA 193 f.). Husserl goes on to assert that 'the authentic presentation and a symbolic one pertaining to it stand in the relation of logical equivalence,' and further that 'all judgments that are connected with the symbolic presentation can afterwards be carried over to [the object] itself' (PA 194).[37]

Now the distinction between 'authentic' and 'symbolic' presentation is explicitly applied by Husserl to what he would later call categorial objects. A multitude, for example, is presented *authentically* if we form an articulate presentation embracing each member in its own right, *symbolically* if we merely take the group in at a 'single glance' (PA 195 ff.). It therefore seems reasonable to say that the Husserl of PA views multitudes as susceptible to being intended both in presence and in absence. We should not of course exaggerate the extent to which the distinctions Husserl worked out only later are found already in PA. We should note in particular that there is no explicit discussion in this work of the 'fulfilment' of an empty (or 'symbolic') act by an intuitive (or 'authentic') one.[38] The notion of fulfilment was introduced only after the publication of PA.[39] Nonetheless Husserl does regard authentic and symbolic acts as capable of being directed toward the same object, even in this early work. The examples given, together with his reference to the 'logical equivalence' of symbolic and authentic presentation, provide ample confirmation of this fact. And since Husserl not only observes the difference between authentic and symbolic presentations, but studies presentations in some detail, we can agree with his later comment that PA represents an 'initial attempt' at a 'phenomenologico-constitutive investigation' of categorial objects. Even in this early work, Husserl is seeking to bring to light the specific set of presences and absences within which categorial objects emerge as identities.

(2) But let us now take up the second of the two questions posed above. Is the problem of the constitutive origin of number in particular addressed in PA? Or does Husserl study only the origin of the *concept* of number in that work? We have seen that number and multitude are, in Husserl's view, closely related.[40] Thus it might be argued that since he studies the constitutive origin of multitude, he studies *a fortiori* the constitutive origin of number. But we must recall that numbers are not simply multitudes; they are rather *determinate* multitudes. And how, we might ask, does this determinateness arise? Does Husserl give a genuinely phenomenological account of the origin of numbers in their distinctness from mere multitudes?

It is difficult to give a straightforward answer to this question. If we confine our attention to the account of the origin of number first presented in BdZ and then repeated prominently in PA, we might well be inclined to answer in the negative. On the other hand, if we examine some departures from and additions to this account made only in PA itself, we could perhaps find indications of a genuinely phenomenological account of the origin of number, i.e. of number as distinct from mere multitude.

Let us look briefly at Husserl's earliest account of the origin of number, the account presented in the final section of BdZ. According to this account, the 'original' numbers are abstract concepts which arise through a certain specification of the abstract concept of multitude. To arrive at these concepts, we must *first* form the general concept of multitude by abstracting it from concrete multitudes (PA 334 f.). This gives us a concept whose content can, according to Husserl, be expressed as follows: 'any something and any something and any something, etc.; or any one and any one and any one, etc.; or, more briefly, one and one and one, etc.' (PA 335). Having abstracted this concept, we are *then* in position to eliminate the 'indeterminateness' expressed by the 'etc.' in the above formulations. 'Through elimination of that indeterminateness, there arise (*entstehen*) from it the concepts of determinate multitude, or *numbers*' (PA 336). Numbers then are held to emerge when we make a certain modification of the abstract concept of multitude. The numbers two, three, four, etc., are in the first instance concepts whose contents could be formulated as follows: "one and one'; 'one and one and one'; 'one and one and one and one'; etc.' (PA 336).

This account, we suggest, should not be regarded as a genuinely phenomenological account of the origin of number. The 'determinateness' which is characteristic of numbers in contrast to mere multitudes is not explained with reference to our experience of concrete groups. It is held rather to arise in a kind of thinking which is directed toward abstract concepts.[41] It is not the presence of concrete groups of objects which is said to lead to number, but certain possibilities for differentiation latent in the abstract concept of multitude.

Now this account is, in its essentials, repeated in Chapter IV of PA (PA 79-82). But in the version presented in PA, we find a brief though important addition. In the very midst of his description of the origin of number through the elimination of the indeterminateness in the abstract concept of multitude, Husserl makes the following comment:

> To derive the number concepts it is not necessary of course to take the general and indeterminate concept of multitude as mediator. We come directly to these concepts from any concrete multitude whatever ... (PA 81).

The suggestion here seems to be that number concepts as such might have their origin in concrete experience just as the concept of multitude in general does. In other words, we could abstract them directly from determinate multitudes of concrete objects. But in that case, the 'determinateness' characteristic of number would have to be present already in the concrete multitudes themselves. The differentiation of two, three, four, etc., would have to take place on the level of our concrete experience of groups, not on the level of abstract, concept-oriented thinking.

It is quite significant that in Chapter V of PA – the first chapter which does not correspond to anything at all in the earlier essay BdZ – we find Husserl going 'back to the circle of concrete phenomena' to study 'the multitude relations of equal, more and fewer' (PA 91, 95). For what we find in this chapter and those that follow it is in fact an attempt to give a phenomenological account of the determinateness by which number is distinguished from multitude in general. Having shown in the previous chapters how 'multitude' is made authentically present, he proceeds in these chapters to show how 'determinate multitude' or number as such is made authentically present. It is only at this point, we suggest, that the authentic counting in which number originates is given a final description.

But it is not just the presence of number that Husserl studies in PA. He also examines, in Part Two of the work, several different modes of absence or 'symbolic presentation' distinctive of the realm of number. Indeed, one entire chapter of Part Two is devoted to 'symbolic number presentations' (PA 222ff.). Thus it seems plausible to say, with reference to PA as a whole, that number is explained in the work with reference to its presence and absence.

PA is by no means a well-organized, finely polished work. Nonetheless it does represent, as Husserl himself later said, an 'initial attempt' to study the constitutive origin of number. Although this fact is obscured by the organization of the work, Husserl does make an attempt in PA to exhibit the structures of presence and absence within which numbers, or determinate multitudes, emerge as identities. This is not of course to imply that the Husserl of PA himself understood what he was doing in just these terms. Nor is it to deny that the account of the origin of number to be found in PA is defective and incomplete in some respects. It can be said, however, that the Husserl of PA had already achieved some important insights regarding the 'phenomena' with which number is involved, or the manner in which number makes its appearance in the world in which we live. In the following chapters of this study, in which we shall examine the origin of number in detail, these insights will be discussed more fully.

NOTES

1 Euclid, *The Thirteen Books of Euclid's Elements*, ed. and tr. by T.L. Heath (New York: Dover, 1956), Vol. II, p. 277.

2 According to Husserl's notes of a lecture delivered on 6 May 1878, Weierstrass described the counting process and then concluded with the remark: 'So entsteht eine Vielheit von Einheiten und dies ist die Zahl' (PA XXIV n.).

3 G. Frege, *The Foundations of Arithmetic*, tr. by J.L. Austin (Oxford: Balckwell, 1968), pp. 79[e] f.

4 In regard to the concept of number, Husserl says, 'there can barely (*kaum*) be talk of definition' (PA 119f.). This statement has often been misinterpreted by commentators. Pietersma, for example, reports that 'the author [of PA] tells us that he does not intend to give a definition of cardinal number' ('Husserl and Frege,' p. 303). Pivcevic goes so far as to say that Husserl 'did not think that the concept of number could be given a logical *definition*' (*Husserl and Phenomenology*, p. 23). But Husserl *does* think number can be defined, as we shall see below. His point is simply that no definition is sufficient to clarify the number concept in a complete manner.

5 PA 15, 81 ff., 336 et passim. According to Husserl's notes, Weierstrass offered a very similar definition in his lecture of 25 October 1880: 'Die Zahl ist also nichts anderes demnach als eine bestimmte Vielheit gleichartiger Dinge' (PA XXIV n.). However, it seems that the term '*bestimmte*' may have a different sense in Weierstrass' definition than it has in Husserl's. For Weierstrass it seems only to signify that we have a 'determinate concept' (*bestimmten Begriff*) of the objects being counted. For Husserl, however, it signifies, as we shall see, that the 'how many' is determinate, not the concept under which the individual counted objects fall.

6 Husserl observes that in a loose sense we sometimes use the term 'number' as a synonym for 'multitude' (PA 14). When we say that 'there were a number of people present,' we may mean simply 'many people.'
7 Frege, *Foundations of Arithmetic*, p. 59[e].
8 We might say that where Husserl rejects only the Euclidean explication of the *arithmos*-concept, Frege rejects the *arithmos*- concept itself. For the ancient Greeks, as J. Klein observes, 'the *arithmos* indicates in each case a *definite number of definite things*. It proclaims that there are precisely so and so many of these things' (*Greek Mathematical Thought*, p. 46). It is this understanding of number that Husserl is seeking to express in his definition; indeed, he seems to express it more clearly than Euclid himself.
9 Frege, *Foundations of Arithmetic*, p. 38[e].
10 Ibid., p. 59[e].
11 In German there are two words for number: *Zahl* and *Anzahl*. *Zahl* is the term used in mathematics 'real number,' for example, is in German '*reele Zahl*.' *Anzahl* on the other hand is the term used in everyday life with reference to counted groups. It is quite similar in meaning to the Greek *arithmos*; see E. Brann's 'Translator's note' in Klein, *Greek Mathematical Thought*, p. vii. *Anzahl* is sometimes translated as 'cardinal number,' *Zahl* simply as 'number.' But in the context of Husserl's philosophy, this seems somewhat inappropriate. It suggests that the mathematical concept of number is the basic one, while *Anzahl* is merely a special case. But Husserl's central point is precisely that the fundamental sense of number is *Anzahl*; mathematical number becomes possible only through a certain modification of the intentional structures at work in our dealings with numbers in the sense of *Anzahlen*.
12 Frege, 'Rezension,' p. 183.
13 Ibid., p. 183 f.
14 'Definitions show their worth by proving fruitful. Those that could just as well be omitted and leave no link missing in the chain of our proofs should be rejected as completely worthless' (Frege, *Foundation of Arithmetic*, p. 81[e].) Cf. Frege, 'Rezension,' p. 184.
15 In his review of PA Frage claims, on the one hand, that only 'psychological logicians' are interested in the senses of words, and, on the other, that the psychological approach, insofar as it pretends to investigate the essence of number, is undermined by the recognition that 'my presentation of number is not number itself' ('Rezension,' pp. 183, 192). Frege is correct in distinguishing numbers themselves both from our presentations of number and from the senses of number words. What he overlooks is the question as to how numbers themselves are available, how there can truly 'be' such a thing as number. As we shall see in the following section, Husserl makes this question of the availability or true being of number the central issue in his philosophical analysis.
16 LU I 244 (238); cf. FTL 189 (181), CM 94 f. (59).
17 T. de Boer makes this point very clearly; see *Development of Husserl's Thought*, pp. 26, 71 f.
18 See PA 15 f., 17, 19, 77 ff.
19 This pattern is most clearly seen in BdZ. In Section 1 (PA 296-302), Husserl states that the concept of number is abstracted from concrete aggregates. In Section 2 and Section 3 (PA 302-334), he describes aggregates in greater detail. Finally, in Section 4 (PA 334-338), he describes the act of abstraction which leads to number. The pattern is not seen as clearly in PA, for in this work Husserl apparently goes 'back to the circle of concrete phenomena' after having described the abstraction of number (PA 91). On the significance of this return to the concrete phenomena, see below, p. 40f.
20 In *Ideen*, for example, Husserl says that 'all acts generally ... are 'objectifying' acts, acts that 'constitute' objects originally (*Gegenstände ursprünglich 'konstituierend'*), the necessary sources of different regions of being and of the ontologies that belong therewith' (*Ideen* 290 [306]).
21 We follow Sokolowski in using the terms 'absence' and 'presence' as alternatives to Husserl's own talk of 'empty' and 'filled' intentions. See Sokolowski, 'The Structure and Content

of Husserl's *Logical Investigations,' Inquiry* 14 (1971), pp. 328 ff.; *Husserlian Meditations*, pp. 18 ff.
22 LU II/2 122, 125 f. (765, 768).
23 E. Tugendhat, *Der Wahrheitsbegriff bei Husserl und Heidegger* (Berlin: Walter de Gruyter, 1967), pp. 174-178. Tugendhat takes the term 'itself-givenness' (*Selbstgegebenheit*) as the clue to Husserl's concept of constitution, and indeed as the fundamental theme of Husserl's philosophy as a whole. And as Tugendhat points out elsewhere in his study, the 'itself' refers to the interplay of empty and filled intentions: 'das 'selbst' in dem jetzigen Sinn (ist) überhaupt nur noch aus der Differenz der *Gegebenheitsweisen* zu verstehen: der Akt der blossen Intention ist auf die Sache *selbst* gerichtet und auf keine andere, aber sie ist ihm nicht *selbst* gegeben. ... die 'Sache selbst' ist das, worin eine objektivierende Intention zur Erfüllung kommt' (ibid., p. 58).
24 LU II/2 153 ff., 160 f. (792 ff., 798 f.).
25 LU I 244, II/1 225, 252, II/2 143 (237, 435, 455, 785); *Ideen* 28, 29 (61, 62); FTL 87 f., 91, 92 (83, 87, 88).
26 LU I 168, 170 (178, 179); FTL 87 f., 91, 112 f. (83, 86 f., 109 f.).
27 It should be observed that Husserl refers to these early studies in the passages in FTL in which 'counting' is described as the activity in which numbers are originally constituted (FTL 88 n., 90 f. [83 n., 86f.]).
28 In 1887 Husserl formally defended the thesis that 'in the authentic sense one can barely count beyond three' (PA 339); cf. Schuhmann, *Husserl-Chronik*, p. 20. We shall reserve for later an evaluation of this claim; for now the important point is merely that Husserl does distinguish between authentic counting and everyday counting. It seems to be authentic counting that is meant at PA 182, where Husserl says that 'numbers come to be through the counting (*Zählung*) of multitudes.' R. Torretti's comment on this passage ('La filosofia de la aritmetica de Husserl,' p. 205) is apt: 'Naturalmente, este procedimiento no puedo identificarse con lo que llamamos *enumeración* (en alemán *Aufzählung* o, también, *Zählung*), y que consiste en determinar el numero de elementos de un conjuncto, pareándolos con un segmento inicial de la serie de los numeros. En el punto en que nos hallamos, aun no se dispone de esa serie, para etablecer el pareo.'
29 As we shall see, everyday counting is an example of what Husserl calls 'sedimented' thinking; see below, pp. 79ff.
30 Biemel, 'The Decisive Phases,' pp. 152, 153; Sokolowski, *Formation of Husserl's Concept of Constitution*, p. 18.
31 De Boer writes that when Husserl speaks of 'origins' in PA, he does not mean an object arising from an act but a concept arising from an object or 'denotatum' (*Begriffsgegenstand*) ...' (*Development of Husserl's Thought*, p. 119; cf. pp. 68 ff.).
32 Ibid., pp. 72, 119 ff.
33 In PA, de Boer claims, 'there is no object of a higher order corresponding to the act of a higher order' (*Development of Husserl's Thought*, p. 24 and 27 n.; pp. 24-29, 119).
34 De Boer is right of course in pointing out (ibid., p. 34) that the Husserl of PA does not investigate the constitutive origins of sensuous or 'lower-order' objects.
35 See above, p. 22f.
36 PA 6f. Part One of the work is entitled 'The Authentic Concepts of Multitude, Unit and Number,' Part Two 'Symbolic Number Concepts and the Logical Sources of the Arithmetic of Numbers.' It should be noted however that the distinction between authentic and symbolic *concepts* is not the same as that between authentic and symbolic *presentations*; on this point see below, p. 78.
37 In a footnote in PA, Husserl expresses a debt to Brentano for having called his attention to the distinction between authentic and symbolic presentation (PA 193 n.). Brentano, he says, 'placed the greatest emphasis' on this distinction in his lectures. We find some confirmation of this in the posthumously published text of Brentano's lecture manuscripts. Brentano observes that inauthentic presentation is at work not only when we speak of God, the infinite or the eternal, but also 'wenn wir Gegenstände nennen, deren einzelne Merkmale wir wohl fassen

können, die aber wegen ihrer Komplikation für uns nicht eigentlich mehr vorstellen und nennen sie, ohne den Namen genau zu verstehen' (*Die Lehre vom richtigen Urteil*), p. 64). Nonetheless it is important to take note of the difference between Brentano's use of this distinction and Husserl's own. Brentano seems content to observe that some objects *are* authentically presentable, while others are not. Husserl on the other hand stresses even in PA that one and the same object can, in some cases at least, be presented *either* authentically *or* symbolically. Landgrebe calls attention to this difference in a general way when he observes: 'Der Gedanke, daß zwei oder mehrere, deskriptiv voneinander verschiedene Akte identisch *dasselbe* intentionale Objekt haben können, der ... für Husserl von Anfang an eine große Rolle spielt, liegt Brentano gänzlich ferne' ('Husserls Phänomenologie und die Motive zu ihrer Umbildung,' p. 11 f.). See Tugendhat, *Wahrheitsbegriff*, p. 31 f.

38 O. Becker seems to be mistaken when he claims to find in PA 'the thought that the merely signitive and empty presentation is to be filled with a self-giving intention ...' ('The Philosophy of Edmund Husserl,' p. 42). The relation asserted in PA to hold between empty and filled acts is the static one of 'logical equivalence,' not the dynamic one of fulfilment.

39 This notion was introduced in Husserl's article 'Psychologische Studien zur elementaren Logik,' *Philosophische Monatshefte* 30 (1894), pp. 159-191. See especially the second study, 'Über Anschauung und Representation,' in which Husserl speaks of acts of intuition as the 'fulfilments' of corresponding acts of mere 'representation.'

40 See above, pp. 32ff.

41 Husserl allows, of course, that once we have determinate number concepts, we can apply them to concrete multitudes (PA 335f.). The point here is only that these concepts do not *arise* in concrete experience according to the doctrine presented in BdZ.

CHAPTER III

THE PRESENCE OF NUMBER

1. SENSUOUS GROUPS

In the previous chapter we saw that, according to Husserl, the concept of number can be clarified by a study of the 'phenomenological origin' of number. Number may be defined as 'determinate multitude,' but we come to a full understanding of what we mean by number, or of what numbers themselves truly are, only through a study of the intentional acts correlated with numbers. In particular, as we saw, Husserl seeks to bring to light the network of empty and filled intentions within which numbers emerge as identities. To study the phenomenological origin or 'constitution' of number is to study the specific modes of presence and absence characteristic of numbers. In this chapter and the one that follows, then, we shall be concerned with the details of Husserl's attempt to clarify number phenomenologically. We shall examine his investigation of the specific modes of presence and absence which are characteristic of the realm of number. In the current chapter we shall be concerned only with that 'authentic counting' which makes number present in an originally intuitive manner. The absence of number will be treated in the chapter that follows.

How are numbers made present originally? In his mature works, Husserl classifies numbers under the broader heading of 'categorial objects.' This classification is meant to bring out the fact that, although numbers are indeed objective, their objectivity is not of the same type as that of sensuous objects. Numbers have a distinctive manner of givenness, a distinctive manner of becoming present. They can never be experienced in straightforward sense perception, as material things can. For numbers to be given authentically, thoughtful, 'categorial' activity of a certain type is essential. Later in the present chapter, we shall attempt to follow Husserl in his precise description of this activity. We shall see how this type of categorial activity leads to the constitution of number. Before describing this activity as such, however, we must examine the context of passivity which it always presupposes.

See notes at the end of this chapter.

In the *Cartesian Meditations* Husserl writes that 'anything built up by activity necessarily presupposes, as the lowest level, a passivity which gives something beforehand; and when we trace anything built actively, we run into constitution by passive generation' (CM 112 [78]). A study of the constitution of any class of categorial objects must begin on the 'lowest level' with a study of the pertinent form of passive givenness. In studying this passive givenness, we acquire both a foundation and a foil for a subsequent description of the active constitution of the categorial objects in question. *Experience and Judgment* offers an example of such a return to the sphere of passive givenness. In this work Husserl is especially concerned with the constitution of facts or 'states-of-affairs' (*Sachverhalte*). Facts are categorial objects, like numbers. They are constituted only through the thoughtful, categorial activity of predicative judging. But to get a complete understanding of the constitution of facts, we must first thematize what Husserl terms the pre-predicative 'situation' (*Sachlage*). A situation, Husserl says, is the 'passively preconstituted foundation' of a fact (EU 288 [241]). When we experience a situation, we have not yet formed an active judgment that 'S is p.' We have not explicitly distinguished *p* from *S* and brought it into prominence as a feature in its own right. But we do experience something that somehow anticipates this active judgment. We do experience the object as vaguely containing something that could be brought to prominence as the feature *p*. When we actively judge that 'S is p,' and thus take the first step toward constituting the fact, we merely actualize a potential for articulation that was there in advance of the judgment itself. It is this passively given potential that Husserl tries to bring to view as a foundation for the constitution of facts.

Numbers and facts are obviously not the same. They are constituted through categorial activities of different types.[1] But it does seem that we should ask whether there is not something which is related to number as the 'situations' discussed in *Experience and Judgment* are related to facts. Is there anything in the sphere of passive givenness which underlies the constitution of number? Is there anything in our receptive experience which we could regard as 'potential numbers,' in the way situations are potential facts? Some of Husserl's remarks seem to suggest a negative answer to these questions.[2] His own analyses, however, clearly show that there is something to be taken account of in studying the constitution of number which is at least quite analogous to the experience of fact-founding situations. In the *Cartesian Meditations*, for example, he describes the phenomenon of 'pairing' (*Paarung*), in which two objects are found to group themselves quite independently of any active collecting or counting on our part (CM 142 [112]). Similar phenomena are described in *Experience and Judgment* itself. Let us then consider these passively constituted groups more closely.

Husserl has such groups in mind when he calls attention in Chapter XI of PA to the following occurrences:

> We enter a hall full of peole; one glance suffices and we judge: a lot of people (*eine Menge Menschen*). We look up to the starry heaven, and in a single glance we judge: many stars (PA 196).

Whenever we take in an entire group of things 'in a single glance,' as in these examples, we are experiencing what Husserl calls a 'sensuous group' (*sinnliche Menge*), a group which 'presents itself ... from the first as unitary intuition, as a whole' (PA 195).[3] In such cases the group we experience is not arbitrarily concocted by us; it is simply encountered. We simply find ourselves in the presence of the group. We 'see' the unitary sensuous group in much the same way that we see a unitary sensuous thing. There is of course a difference between the unity of a thing and the unity of a sensuous group. What is distinctive of a sensuous group is the peculiar counterpoise of unity and discreteness that is at work. Each item of the group appears to have a certain integrity and independence, yet at the same time all the items seem to 'belong together' with one another. Both discreteness and unity are, it seems, essential to the phenomenon of sensuous groups. If either disappears, we no longer find ourselves experiencing a sensuous group. If we imagine the discreteness of the items to be exaggerated (their distance from one another to be increased, for example), there comes a point where the unity vanishes and we find ourselves unable to continue seeing the group. And likewise, if we imagine the unity to be exaggerated, there comes a point where the discreteness disappears and where, for this reason, we also cease to see a group. We may 'know' in such a case that what we see is many; but we do not 'see' the group as we do in other cases.[4]

As Husserl points out in PA, a certain experience of 'likeness' is involved whenever a sensuous group comes before us (PA 208 ff.). This is not of course a distinctly apprehended likeness of the sort which might arise if we actively compared the items of the group. It is rather a vague similarity which we are aware of as somehow permeating an entire group of objects. The similarity might be one of shape or size or color; or it might merely be that all the items are moving in the same direction (cf. PA 205 f.). But in any case we perceive the similarity in a passive way, without need of active, explicitly comparative thought. The similarity we experience 'gives to the groupwhole a specific character which comes to the fore without every individual thing having been compared with every other one' (PA 208).

> The individual content hangs, as it were, on a chain which it pulls along behind it as soon as it comes into consciousness. If we attend, for example, to a white square on a chess board, the whole configuration of white squares emerges; if we attend to a black square, then that of the black squares (PA 213).

Whenever we experience a sensuous group, some such passively given likeness or similarity is at work.

Another factor in our experience of sensuous groups is the 'configuration' of the items (PA 205). In some cases we find the items of the group to have a 'linear' arrangement, in others a 'triangular' one, in others perhaps a 'pentagonal' one. Many other, less well-defined configurations are also possible, of course. But in order to experience such a configuration, we need not have 'analyzed' the group into its components. A 'single glance' is all that is needed (PA 204, 205). The configuration 'appears to us immediately, and only in subsequent reflection do we notice the founding ... conditions' (PA 205). If these 'founding conditions' should change, what we notice first of all is a change in the configuration of the group as a whole. For example, if the relative positions of the items in a linearly configured sensuous group are altered, we might notice the change in the total configuration even before noticing the changes that brought this about. 'Every variation of positional relations brings about a variation of figure, and vice versa; but we notice the variation of figure before becoming aware that these or those relations were changed' (PA 205; cf. 208, 209). A single, unarticulated glance suffices to tell us that the group is now configured differently.

We might well ask, as Husserl himself does, how it is possible for us to experience such sensuous groups. How is it possible for us to take in an entire group of objects 'in a single glance'? There is one account of the presence of sensuous groups which Husserl is particularly concerned to avoid. It is not the case, he writes in PA, that we are 'unconsciously' running through the items of the group and collecting them in active thought (PA 196 f., 199 f.). We are certainly not aware of any such 'lightning-fast sequence of separate apprehensions and connections,' and it is implausible to suppose that an 'extensive and animated activity' of this sort would be forgotten right away (PA 197). The experience of sensuous groups must be accounted for in some manner that is compatible with its manifest passivity.

In PA Husserl explains the experience of sensuous groups with reference to what he terms the 'figural' or 'quasi-qualitative' moments of such groups. The pervasive similarity we have described above is one such moment or feature; the configuration of the items is another.[5] When we experience a sensuous group, Husserl claims, we are in fact taking such a moment as a 'sign' (*Anzeichen*) of the group as such. 'In the intuition of the sensuous group there must lie an *immediately graspable sign* through which the group character is known' (PA 201). The moment itself is experienced in a straightforward, sensuous manner, then taken to indicate that what we are experiencing is in fact a sensuous group.[6]

In LU Husserl evidently still adheres to this account of the experience of sensuous groups. What he had earlier called 'figural' of 'quasi-qualitative' moments are in this work called 'moments of unity' (LU II/1 234 f. [442]). And Husserl claims in LU, just as he had in PA, that such moments are what underlie the passively encountered phenomenon of sensuous groups (LU II/1 243 [449]). These moments serve as 'signs of plurality,' as 'sensuous *points d'appui* for the signitively mediated cognition of plurality as such' (LU II/2 161 [799]).

In later writings, however, we find a somewhat different account of the experience of sensuous groups. Husserl does not of course deny in his later works that sensuous groups have what he had earlier called figural moments (cf. EU 153 ff. [134 ff.]). But he does cease to assert that in the experience of sensuous groups, figural moments serve as 'signs' of the groups themselves. In his later works this experience is explained not as a case of sign consciousness, but rather as the result of what is termed the 'passive synthesis' of sense data. Sense data, Husserl claims, have a way of arranging themselves in the visual field even before they are apprehended. Two or more sense data can come to prominence together, but in such a way that their discreteness is not destroyed. One datum that has come to prominence can bring others trailing along behind it, so to speak. This is the basis of the phenomenon which is referred to in the *Cartesian Meditations* as 'pairing.' Seen in the most elementary case, the essential feature of pairing is that 'two data are given intuitionally, and with prominence, in the unity of a consciousness and that, on this basis ... they *found phenomenologically a unity of similarity* and thus are always constitued precisely as a pair' (CM 142 [112]). Husserl stresses that the data in question come to prominence together 'already in pure passivity (regardless therefore of whether they are noticed or unnoticed)' (CM 142 [112]). The similarity which permeates the two data and makes them into a unified prominence in the sensory field is not something we grasp through an explicit act of comparison. On the level of consciousness currently under discussion, there is no thoughtful asserting that 'A is like B.' It is not that objects are being compared, but rather that the sense data are pairing themselves. This phenomenon of 'paring,' of course, sometimes involves more than two data (cf. CM 142 [112]). There will sometimes be several discrete sense data which come to prominence together. The several data have a certain 'resonance' with one another and form, already on the level of passive consciousness, a kind of 'community' among themselves.[7] For example, if one red patch stands out from a predominantly white background, it will tend to associate to itself any other red patches which may be in the visual field (EU 76 f. [73 f.]). In virtue of the as yet unthematized similarity which permeates them, the red patches drift into prominence one after the other, forming a unified, peculiarly 'fused' prominence that stands out as a whole from the remainder of the visual field.[8]

This passive synthesis of sense data is not itself the objective unity of a sensuous group, of course. For us actually to experience an objective group, for the group itself actually to appear to us, the sense data must be 'apprehended' in the appropriate manner. We must actually 'turn toward' the sense data, as in other cases of sense perception.[9] What is distinctive in the case of sensuous groups is that the sense data exert a kind of collective pull even before they are apprehended. They 'clamor' for attention as a single whole.

> Even before all turning-toward, the individual data affecting [the ego] in discreteness have a community; they stand in a relation of 'resonance.' One raises the other, i.e. the affective stimulus which one has on the ego raises the affective stimulus of the other, and vice versa – but in such a way that these affective stimuli do not remain apart, but rather come together in the unity of many-rayed affective stimulus in which the increased affective stimuli are united ...[10]

This union of sense data can then be apprehended in a single thematizing glance. The ego 'turns toward' the sense data all at once. Thus it is 'on the basis of a unified prominence and affection' that 'a plurality ... [becomes] a unitary theme ...' (EU 134 [120]). The passively given affective unity receives an objectivating interpretation; and as this happens, as the ego turns toward the passive unity of sense data as a whole, 'it grasps the sensuous group in an unsevered way.'[11] We do not first turn toward or apprehend each sense datum separately, then deliberately unite the objects perceived into a group. Rather, a single glance suffices to apprehend the group as a whole. Because the data have already grouped themselves prior to all ego activity, there is no need for a series of glances. Perceiving a sensuous group is a matter of seeing it in a straightforward manner.

This account of the experience of sensuous groups seems preferable to that which Husserl had given in PA and LU. Certainly Husserl was right to stress in PA that our experience of sensuous groups cannot be described in terms of unconscious mental activity. But he seems to have been somewhat too hasty to assume that the experience of sensuous groups involves instead the taking of one thing (the figural moment) as a 'sign' of another (the sensuous group as such).[12] It is difficult to see how the configuration by itself could ever be said to serve as a sign 'through which the group character can be known' (PA 201). In some situations, we might indeed take the features of an object as a sign of the object. For example, we might take the chemical properties of an unknown body as a sign that the body in question is aluminum (cf. PA 304 f.). But for this inference to be plausible, the chemical properties in question would have to be true characteristics of aluminum. A moment of configuration, however, is not truly characteristic of a sensuous group. A material thing and sensuous group of three items might both exhibit a 'triangular' figure; they would have the same figural moment, even though the former obviously would not be a sensuous group at all. There is also another difficulty with the account of sensuous groups presented in PA and LU. By saying that we first experience the figural moment, then take it as a sign of the sensuous group, Husserl seems to be suggesting that the sensuous group itself is not actually present to us at all, that it is merely inferred on the basis of the figural moment.[13] But surely this is incorrect. When we 'see' a group in a single glance, it is precisely the group itself that we see. The group itself is present to us, not something else which allows us to infer the group character of what we see.

Husserl's later account of the experience of sensuous groups is preferable to his earlier account because it is more compatible with the directness and immediacy of this experience. The later account stresses that what we see when we experience a sensuous group is the group itself, not something else which serves as a sign of the group. We see the sensuous group itself, just as we might see a material thing. In PA Husserl had seemed to deny this, but he had at least introduced the topic of sensuous groups. This in itself is quite important. Even though the presence of a sensuous group is not as such a presence of number, it is nonetheless an essential component of the presence of number.[14] Such groups provide the sensuous foundation for the 'authentic counting' which generates an original presence of number. They underly the original determination of 'how many' that leads to a constitution of number. In the following sections, we shall examine the categorial activities through which this determination takes place in more detail.

2. EXPLICATION

Husserl distinguishes – in PA and in later works as well – between the straightforward presentation of a sensuous group and the complex, articulated act in which we apprehend each item or member of the group in its own right. A sensuous group is characteristically presented, as we have seen, 'in a single glance.' But over against this instantaneous group presentation stands the possibility of 'running through' the items of the group one-by-one, taking note of each in turn and thereby achieving what Husserl terms a 'successive apprehension' of all the members of the group (PA 198). This more active registration – let us refer to it as 'taking stock' – makes the sensuous group in question present in a new manner. If we are to see how numbers are made present originally, we must discuss the activity of stock-taking in more detail.

The articulated act which we are seeking to describe is an instance of what Husserl's calls 'explication.'[15] That is, it is an act in which the parts of an at first indiscriminately given whole are brought to explicit prominence as parts of the whole in question (EU 112 ff. [103 ff]). Now we may note first of all the fact that the stock-taking explication of a sensuous group is not a wholly arbitrary activity. In taking stock, we respond to something present in the sensous group itself. We do not always actively explicate the groups we confront, of course. But when we do, we actualize a possibility that is there in advance of our own activity, a possibility that is latent in the sensuous group itself.[16]

What sort of explication is involved in taking stock of a sensuous group? In general, Husserl distinguishes between two kinds of explication and, correlatively, between two kinds of parts engendered through explicative activity.[17] There is, on the one hand, the kind of explication which leads to a presence of the non-independent parts or 'moments' of a thing. When we analyze a presented thing so as to allow it to become present as a 'substrate' having various 'properties,' we are explicating it in this way. Properties are 'non-independent' parts because they cannot be separated from the substrates in which they inhere; any intuitive presentation of a property, even an imaginary one, requires that its substrate be presented also. The case is otherwise, however, with independent parts or 'pieces' (*Stücke*). Pieces, Husserl says, can by their very nature be presented separately from the wholes to which they belong. A horse's head, for example, is said to be a 'piece' of the horse as a whole, since it 'can be presented 'on its own' or 'cut off,' i.e. we can hold it in our fancy, while we allow the other parts of the horse, and its whole intuited setting to vanish at will' (LU II/1 231 [439]). Pieces can be presented apart from the wholes to which they belong, at least in imagination, and thus must be distinguished from moments, which cannot. Likewise, the explicative activity which engenders pieces – Husserl calls it 'fragmentation' (*Zerstückung*) – must be distinguished from the explicative actity which engenders moments.

The kind of explication with which we are now concerned – that in which we take stock of a sensuous group by 'running through' the items one-by-one – clearly has more in common with 'fragmentation' than it does with analysis of an object into a substrate with properties. The items of the sensuous group come to be presented as 'independent' parts of the group as a whole, parts which could also be presented 'on their own,' apart from the group to which they belong. To be sure, sensuous groups can have moments or properties, just as real, sensuous individuals can. Groups can emerge as substrates in which certain features, e.g. the 'figural moments' discussed in the previous section, inhere.[18] But the items comprised in a group are not themselves features of the group. 'In a sensuous group,' as Husserl writes in PA, 'the parts are not contained in the manner of properties, but rather in the manner of *discrete part intuitions* in their own rights' (PA 195). The phrase 'discrete part-intuitions' is being used here to designate the same phenomena that Husserl would later call 'independent parts.' Thus he is pointing out in effect that the items of a sensuous group are independent parts, that the explicating activity through which they come to prominence is not of the kind that engenders moments.

The stock-taking explication of sensuous groups, then, is a form of explication that engenders independent parts, not dependent parts. But it seems that a further distinction can be made here. There is a difference between the independent parts of a real individual and the independent parts of a sensuous group, and the form of explication in which the latter emerge would seem to differ from that in which the former do so. In PA, where Husserl sometimes uses the term 'physical parts' to designate what he would later call 'pieces' (cf. PA 72), the situation is described as follows:

> The presentation of physical parts can enter into a thing presentation as well; but here our attention rests on the connection of those parts with the whole, on the fact that they belong to it – which makes them *characteristics (Merkmale)* of the whole. Presentations of physical parts belong to a group presentation in a quite different manner. Each one stands on its own, not as a characteristic of the whole. Here the parts are intuitively discrete, so that their connection recedes within the intuition of the whole. Moreover, where a sharp separation of the part-intuitions is lacking, it depends on the direction of our interest whether we speak of a group or of an intuitive thing (a tree or a group of branches) (PA 195 n.).

Husserl seems to be suggesting here that the pieces or physical parts of a material thing occupy an intermediate status in regard to their independence from the wholes to which they belong. They are not dependent parts in the manner of properties, but neither are they fully independent parts in the manner of a sensuous group. It is only in the latter case that we find parts which can actually stand on their own, parts whose connection with their wholes can truly vanish altogether. The items of a sensuous group, it would seem, are more genuinely independent of the wholes to which they belong than are the pieces of a material thing.[19]

The stock-taking explication of sensuous groups, in which items become distinguished as truly independent parts, can be described more precisely. We may observe, for example, that it always has the character of a 'process,' that is, of a temporally extended sequence of acts. We cannot take stock of a sensuous group in a single, instantaneous act. We cannot focus directly upon and scrutinize several objects simultaneously. There must be one intentional act for each item in the group, and these must be performed in succession to one another. But even though we can apprehend only one item at a time, at each stage in the process we nonetheless *anticipate* apprehending the other items as well. The other items in the group are there for us as something that we could 'also' take stock of, even though we are not engaged in doing so at present. We scrutinize one soldier in the column, and as we do so each of the others is somehow there as something that we could, in turn, 'also' scrutinize. These others are there for us, moreover, with more prominence than that enjoyed by mere background figures. They are there as thematizable individuals which 'belong together' with the one currently under scrunity. Husserl describes this situation in *Experience and Judgment* with reference to a whole consisting of two independent parts. As we focus on the one 'piece,' it becomes raised into the explicit prominence of the center of the sensory field. But this is not all that happens, for it is also the case, he says, that 'an *excess*, a surplus, comes into prominence, a surplus which has its own affective force and is apprehensible as a second piece *connected* with the first' (EU 164 [143]). That is to say, there appears, in addition to the first piece, also an as yet unexplicated 'remainder,' another piece which we could focus upon even though we do not do so at present. Furthermore, Husserls claims,

> the way in which the nonexplicated 'remainder' is present to consciousness is completely different in explication by pieces than in explication by dependent moments ... If a piece is brought into relief, then the nonexplicated 'remainder' is '*exterior*' to it and prominent in relation to it, even though connected with it (EU 164 [143]).

Although Husserl makes explicit reference only to 'pieces' here, what he says would seem to apply to 'items' as well. Even though the items of a sensuous group do not 'interpenetrate' one another – as the moments of color and extension in a physical thing do, for example – they are nonetheless experienced as belonging together. They belong together in such a way that, as we focus on one item, we anticipate focusing on the others in turn. This phenomenon can be accounted for phenomenologically in terms of 'passive synthesis.' The sense data on which the apprehension of the sensuous group–whole is based are themselves discrete, though passively synthesized in virtue of their 'similarity.' We must think of this passive synthesis as continuing, and continuing to exert its 'pull' on the ego, even while the explication goes on. The passive synthesis of discrete sense data does not cease when we begin to explicate, but rather continues to serve as the passive matrix within which active, explicating thought takes place.

Another aspect of the stock-taking explication of sensuous groups emerges if we consider the fact that previous acts of apprehension are '*retained*' in consciousness even as we move forward to further acts. As subsequent items come to prominence, those which were previously prominent do not immediately lapse back into obscurity. They are still there for us with a certain distinctness, precisely as items of which stock has been taken.[20] This retained presence of the already explicated portion of the group is thus of a different character from the obscure, merely anticipatory presence of the as yet unexplicated portion. There is, of course a limit on our ability to retain previous acts with distinctness. As Husserl notes in PA:

> Only a small number [of the successively occurring apprehensions of the members of a group] remain at any given moment in the realm of distinct collecting activity. As new members are apprehended and added on, others already singled out fall away; the acts in which they are presented in themselves become blurred in the background of consciousness, and finally they disappear altogether (PA 198).

When previous apprehending acts are no longer retained with distinctness, the objects at which they were directed come to be intended with a certain obscurity. That portion of the group, though still perhaps sensuously present to us, has ceased to be present in articulate, categorial fashion. The obscurity of a previously explicated but no longer retained portion of a sensuous group is nonetheless different from the obscurity of an as yet unexplicated portion. Even where we have ceased to retain previous acts with distinctness, 'we nonetheless have a knowledge that the [presently retained] portion is not the whole process' (PA 198). 'By way of the association of ideas,' Husserl says, 'we are led back along the chain to the earlier steps, or at least to the memory that earlier steps had been undertaken' (PA 199). In other words, though we may not actually enjoy distinctly present apprehensions of the previously explicated items, we are nonetheless conscious of them as 'previously explicated.' The obscure presence of something that has been explicated is not the same as the obscure presence of something that is yet to be explicated for the first time.

The stock-taking explication of a sensuous group serves, despite the limitations on our retentional ability, to make the sensuous group present to us in a new manner. The sensuous group is no longer the sheer, undifferentiated presence which the original 'single glance' disclosed. It has come to be present in a new, more articulate manner, as a whole comprising distinct parts which have been run through one-by-one. In Husserl's terms, the sensuous group has come to be present as a distinctly collected 'aggregate' or 'multitude.' Correlatively, each of the items is now present precisely as a 'one' (PA 79 f., 84). The difference between the articulated presence of an aggregate or multitude and the initial, as yet unexplicated presence of the sensuous group is quite important. We cannot understand the origin of number unless we are clear about it and the categorial activity that engenders it. But another type of categorial activity is also essential to the constitution of number. It is to this further activity that we now turn.

3. COMPARISON

When we explicate a sensuous group by running through the items it contains one-by-one, we make the group present in a new manner. The sensuous group is present as a distinctly collected multitude or aggregate. But the categorial activity which engenders multitudes is not by itself sufficient to engender a presence of number. For numbers are not merely multitudes; they are rather *determinate* multitudes.[21] A further categorial activity, in addition to that which we have called stock-taking explication, is necessary if numbers in their full determinateness are to be made present in an original and authentic manner. What is this further categorial activity? It is an activity in which we *compare* multitudes with each other. When we compare a distinctly presented multitude with another multitude, the relation of equality or inequality in which they stand comes to be disclosed. Without these relations and the comparing activity which engenders them, we do not have a presence of number.[22]

It seems that Husserl did not at first appreciate the importance of this comparing activity. It is not mentioned at all in the account of the origin of number given in BdZ.[23] It is barely alluded to in the corresponding sections of PA.[24] Later chapters of PA suggest however that Husserl eventually came to attach considerable significance to this activity. In Chapter V of PA, we find the following comment:

> The multitude relations of equal, more and fewer condition the origin of the number concepts in an essential way ... In order for us to move beyond the concept of a 'multitude of units' and form the series of numbers two, three, etc., we must classify multitudes of units, and this requires judgments about equality and inequality. ... (PA 95).

Numbers as distinct from mere multitudes become present to us, in other words, only when we apprehend the multitudes as standing to each other in the relations of more, fewer and equal. The comparing activity which generates these relations is therefore essential to the origin of number.[25]

It is important to note that equality and inequality are relations that hold between multitudes, not between sensuous groups as such. Sensuous groups can of course be compared in certain respects. We might well find that two sensuous groups are 'like' one another in regard to the figural moments they possess or even in regard to the amounts they contain. But a 'likeness' of this sort is not yet 'equality.'[26] A sensuous group as such cannot enter into a relation of equality. For equality to emerge, we must actively explicate the groups in question and allow each item to be given in its own right. It is only when we move from sensuous groups to multitudes that the relation of equality – and likewise, of course, the corresponding relation of the inequality – can be constituted.

Let us seek to describe the comparing act which generates the relations of equality and inequality more precisely.[27] According to Husserl, 'two collective wholes are compared when one seeks to bring them element for element to mutual *coincidence*' (PA 102). If they can be brought to such coincidence, then the two multitudes are 'equal'; if not, the one is 'more,' the other 'fewer.' But what is the mutual coincidence which Husserl speaks of here? Its character is made somewhat clearer by the following passage:

> If we think of the act of comparison as a successive process, as it usually is, then:
>
> 1) to each element of the group with which we begin there must correspond a like one (*ein gleiches*) in the other group, and indeed a new one at each step.
>
> 2) In the second group no elements must be left over. ...
>
> Only if these conditions are fulfilled can we say that the two multitudes as such are equal (*gleich*) (PA 102).

Now this description is somewhat puzzling at first. What does Husserl mean by saying that to each element in the first group there must correspond a 'like' element in the second? Does he mean perhaps that we can only compare multitudes whose elements are somehow 'like' one another? Apparently he does not. Husserl does mention the special case where the items of the multitudes to be compared are all of the same genus (PA 103). He also mentions the case where a multitude is compared to itself at a different time (PA 91). In these cases there is obviously a genuine likeness in regard to content between the items of the one multitude and the items of the other.[28] But Husserl evidently does not consider such likeness to be an essential presupposition of the comparing activity. He points out that we can compare two multitudes even if they consist of 'heterogeneous contents' (PA 93). For 'when each concretely present content is considered only insofar as it is a something, and thought together with the others, then each has become like to each – precisely as a something' (PA 94). What this seems to mean is simply that the items or 'ones' of any given multitude are in a sense 'like' those of any other. All are alike in being elements of collective wholes. We may therefore compare any two multitudes whatsoever.

It should be noted that in the passage quoted above (from PA 102), Husserl says that comparison is 'usually' (*in der Regel*) a 'successive process.' He subsequently explains that this is not always the case: 'With groups having few elements (perhaps two to four), a knowledge of the equality of the collections as wholes can, under especially favorable conditions, become evident in a single glance, so to speak, without successive individual comparisons being necessary' (PA 102). This statement is somewhat misleading, for it suggests that the relation of equality could obtain between sensuous groups after all. It suggests that we need not always take stock of the groups in question in order to register that they are in fact 'equal.'[29] But we have noted previously that this is not in fact the case. Sensuous groups can be 'like' one another, but only distinctly apprehended multitudes can be 'equal.'

There is however a sense in which the comparison of small multitudes does differ from the comparison of large ones. As we saw in the previous section of this chapter, with small multitudes it is possible to retain all of the explicating acts, with full distinctness, in a single moment of consciousness. With large multitudes, however, this is no longer the case; here the previously explicated portion falls into obscurity as we move forward in the explicating process. This difference on the level of explication leads to a corresponding difference on the level of comparison. Let us follow Husserl in examining this latter difference more closely.

Husserl observes in Chapter V of PA that 'we have the ability to make several aggregates present together, as united in a *single* aggregate, without them thereby losing their peculiar unities' (PA 92). From the context it is clear that he is referring to cases where the aggregates or multitudes in question are quite small ones. He gives the following example: 'If I broaden ... the aggregate (A, B, C) to (A, B, C, D, E), then the judgment that the second aggregate is more by D and E requires a simultaneous presentation of (A, B, C), (A, B, D, D, E) and (A,B,C; D, E) – all in a *single* act' (PA 92). In this example, of course, the two multitudes to be compared are really 'the same' multitude at two different times. But Husserl's point here is that, at least in a case such as this, we can enjoy a simultaneous presence of each of the two multitudes in question. Both multitudes are given with full distinctness in a single moment of consciousness; in neither case have the apprehending acts begun to lapse into obscurity. And where this simultaneous co-presence of two multitudes occurs, the act of comparing them can take on a uniquely intuitive character. We can register in a direct and artless manner that the two multitudes either do or do not coincide with each other, that the one multitude either does or does not contain a 'surplus' as compared with the other.

According to Husserl, it is only when we actively compare two small multitudes presented in this manner that we can have an 'authentic presentation' of number. This comparing activity is what makes a determinate 'how many,' as opposed to a mere 'many,' present in an original and authentic manner. Of course, it is not sufficient merely to register that the one multitude is 'equal' or 'unequal' to the other. As Husserl says at one point in PA, an 'exact judgment' of equality or inequality is required (PA 95). A multitude of four and a multitude of three are both 'more' than a multitude of two. Simply registering that they are more does not make them present as fully determinate multitudes or numbers. We must register not only that each contains a surplus, but also that the surpluses in question are themselves related as more and fewer. A multitude of three, we must come to see, differs from a multitude of four as well as from a multitude of two. Only when such 'exact judgments' of equality and inequality are made explicitly, and in the authentic presence of the multitudes in question, do the latter come to be presented as numbers in the specific sense.

It is obviously of course that not all multitudes can be compared in the way multitudes of two, three and four can be. As Husserl puts it, 'the limits for authentic presentation are encountered soon enough, and verything else is only inauthentic (symbolic) presentation' (PA 92). Where the explicated groups are too large to allow a simultaneous retention of all the explicating acts, the artless comparison described above cannot take place. Rather, if comparison is to occur at all in such cases, we must, it would seem, find a relation which makes it possible to set the items of the one multitude into 'one-to-one correspondence' with the items of the other. We must methodically correlate the items of the two multitudes with each other. 'If we compare a heap of apples with a heap of nuts, in order to ascertain whether equally many are present in each ...,' for example, it would seem necessary to 'lay one apple and one nut together repeatedly' until we had exhausted either one multitude or both of them (PA 107 f.). Such methodical procedures can counteract the limitations on our ability to retain the apprehending acts. They make it possible, it would seem, to register equalities and inequalities that would otherwise elude us.

Husserl's own views on the nature and significance of one-to-one correspondence are rather difficult to ascertain. His discussion of this matter in Chapters VI and VII of PA is primarily critical in nature. He cites the views of several theorists (including Frege) who attach great importance to this notion, and then proceeds to dispute many of their claims. In Chapter VI he is concerned with the view that 'equality' or 'equivalence' can actually be defined with reference to one-to-one correspondence. In Chapter VII he is concerned with the further view that, once equality is defined in this manner, number itself can be defined in turn through equality. Let us examine Husserl's criticisms of these views.

In Chapter VI of PA, Husserl cites and criticizes Stolz' claim that 'two multitudes are said to be *equal* to each other ... if with each thing of the first multitude a thing in the second can be coordinated and if none of these remains unconnected' (PA 98). Husserl denies that this can stand as an explication of what we actually mean by the term 'equal': 'by this definition something obvious and well-known in itself is obscured by something remote and foreign' (PA 99). 'It could indeed happen that, in order to ascertain *in concreto* the equality of two groups in regard to their multitude, we lay the elements in pairs or connect them with each other in some other way; but we do not consider this operation to be necessary in all cases, and even where it does take place, the essence of the comparing act does not lie in it alone' (PA 99). Where we are dealing with very small multitudes, it is not necessary to specify a coordinating relation between the items of the two multitudes. And even in cases where we do specify such a relation, the essential thing is not the relation itself, but rather the fact that 'the corresponding elements be connected in our thinking, that they be collected' (PA 109). Merely placing the apples and nuts in pairs does not by itself establish a relation of equality or inequality; it is only in a thoughtful act which runs through the pairs one-by-one, and takes them as pairs of elements from compared groups, that these relations come to be established (PA 108 f.). Thus while it can indeed be said that one-to-one correspondence is a reliable *criterion* of equality, these notions cannot by any means be *identified* with each other (PA 105).

The distinction which Husserl is making here between the relation itself and the criterion by which the relation can be recognized is a plausible one. But Husserl does more in PA then merely to distinguish between equality and one-to-one correspondence. He also claims that the latter is virtually 'useless' as a criterion of the former. It is useful, he contends, 'only for that mental level on which the classification of multitudes (the distinguishing and naming of numbers), and the mechanical counting procedure that is based on it, is still undeveloped' (PA 105). For those 'savages' who find it difficult to count beyond five or so, one-to-one correspondences may be useful; but where mechanical counting is at our disposal, the simples way to compare two multitudes is obviously to count first the one, then the other, and then to compare the results (PA 104 f.). Now this observation is undoubtedly correct, but it is somewhat puzzling in the context. For in fact, the question at issue here is precisely that regarding the 'origin' of number. We are trying to bring out the intentional structures which underly mechanical counting, to analyze that 'authentic counting' which all mechanical counting presupposes. And it might well be the case that one-to-one correspondence plays an important role in that original 'classification of multitude' which is an essential element in authentic counting. For how, we might wonder, could a large multitude be presented originally as a number, as a determinate 'how many' rather than a mere 'many,' if we could not establish some much correspondence?

Husserl seems to have altered his position on the role of one-to-one correspondences after the publication of PA. In Chapter VII of that work, he had claimed that one-to-one correspondence is a notion that is 'fully useless and indifferent' in connection with the problem of the origin and sense of the number concept (PA 117). But in a manuscript apparently written shortly after the publication of PA, this earlier claim is disputed:

> In my chapter on number definitions through equivalence, I was thus decidedly in error. There of course I was dealing with authentic number presentations, and for them such definitions are not needed. But I overlooked the fact that the inauthentic number presentations require a principle of classification that is not given *a priori*, so that we cannot know *a priori* whether a classification is at all possible. The equivalence concept serves this purpose in an outstanding manner.[30]

What Husserl is admitting here is that one-to-one correspondence or 'equivalence' does in fact play an important role in the emergence of large numbers. Of course, neither number nor equality can actually be 'defined' in terms of one-to-one correspondence, for such correspondence plays no role in the authentic presentation of small numbers. But where large multitudes are in question, one-to-one correondences are essential. It is only because we can establish such correspondences that large multitudes can be classified as 'equal' or 'unequal' at all. Without them, we could never make the transition in such cases from the presence of a 'many' to the presence of a determinate 'how many'; we could never move from multitude to number itself.

We may conclude this chapter with a brief summary of the elements at work in the original presence of number. There are three such elements: (1) the presence of a sensuous group; (2) a distinct explicating or collecting of the group's items; and (3) a distinct comparing of the multitude that is generated through such explication with other multitudes. It is only where these elements are at work together that we can speak of an original presence of number. Elements (2) and (3) are categorial acts whose character is different in regard to large groups from what it is in regard to small ones. But if number is to be made present at all, these acts must be performed, and performed in a fully explicit manner. Only through such performances does a sensuous group come to be given as a 'multitude' and eventually as a 'determinate multitude' or number.

Distinguishing and analyzing these elements enables us to understand the original presence of number. But it does not by itself amount to a full account of the 'origin' of number. Number is not just a presence; what we mean by number is rather something that is the same in certain distinctive modes of presence and absence. To give a full phenomenological account of the origin of number, it is therefore necessary that we also examine the relevant modes of absence. It is to this further dimension of the problem that we shall turn in the chapter that follows.

NOTES

1 In FTL Husserl stresses that judging and counting are *related* activities (cf. FTL 112 f. [107 f.]); but this is not of course to deny their difference.

2 In EU Husserl writes at one point that 'there are ... no originally passively preconstituted sets' (EU 296 [247]).

3 Husserl remarks that the term '*sinnliche Menge*' is 'not entirely correct' (PA 195). This is because he elsewhere uses the term '*Menge*,' or 'set,' as equivalent to the terms 'multitude' and 'aggregate' (cf. PA 14, 139); and his contention is precisely that multitudes or aggregates or sets cannot be presented 'sensuously.' In the present study, we shall avoid this terminological difficulty by speaking of 'sensuous groups' rather than 'sensuous sets.' – Although the examples cited above from PA 196 might seem to suggest that all sensuous groups are groups which contain a great many items, Husserl clearly intends the term to apply to small groups as well (cf. PA 216).

4 Husserl finds an indication of the phenomenon of sensuous groups in 'the language of ordinary life': 'One speaks for example of a column of soldiers, a pile of apples, an avenue of trees, a covey of fowl, a flock of birds, a flight of geese, etc. In each of the examples there is reference to a sensuous group of objects similar among themselves, objects whose genus is also named' (PA 203 f.). Another linguistic indication of this phenomenon might be found in the plural and dual forms of nouns and pronouns that are found in some languages. On this point, see W.J. Entwistle, *Aspects of Language* (London: Faber and Faber, 1953), p. 196: 'The plural would seem to originate not in enumeration but in savages' incapacity for number. At some point their powers of reckoning give out and they speak of an indeterminate multiplicity. Above *one* there are well-known groups of doublets, such as eyes, feet, legs, parents, etc. These offer support for the dual number, applied to things in pairs.'

5 Husserl often uses the term 'figural moment' in PA not only for the configuration itself, but also for the moment of 'likeness' present in a sensuous group. At PA 211, for example, he refers to 'figural moments of both kinds (for example, the configuration in the narrower sense in fusion with the moment of likeness or quality) ...' In this broad sense, 'figural moment' is synonymous with 'quasi-qualitative moment.' This may seem a peculiar usage, but Husserl justifies it on the ground that configuration is 'the most notable special case' of the phenomenon in question (PA 209).

6 In a footnote Husserl acknowledges that what he terms 'quasi-qualitative' or 'figural' moments are basically the same as what Christian von Ehrenfehls had investigated in his 1890 article 'Über Gestaltqualitäten.' Husserl denies having been influenced by von Ehrenfehls, however, and suggests instead that the seeds of his ideas were planted by E. Mach's *Beiträge zur Analyse der Empfindungen*, a source which von Ehrenfehls also acknowledges. – The admitted kinship between Husserl's quasi-qualitative moments and von Ehrenfehl's *Gestalt*-quali-

ties suggests the broader question of the relationship between Husserl's work in this area and that of *Gestalt*-psychologists such as Wertheimer, Koehler and Koffka, who seem to have taken their initial inspiration from the von Ehrenfehls article. A. Gurwitsch has proposed the thesis that there is 'a certain convergence between the general orientation of phenomenology and that of Gestalt theory' (*The Field of Consciousness* (Pittsburgh: Duquesne University Press, 1964), p. 7). He shows many respects in which the two disciplines overlap and complement each other, especially in their descriptions of perceptual consciousness. There would seem however to be points of divergence as well. The Gestalt theorists are primarily psychologists. When they go beyond the descriptive study of consciousness, it is often to formulate theories about the physiological basis of consciousness. Thus they have asserted the so-called principle of 'isomorphism,' which states that experiences always have the same structures as the brain processes which underly them. It scarcely needs to be said that theorizing of this sort is quite foreign to Husserl himself.

7 *Husserliana*, Vol. XI, p. 418.

8 Husserl had said already in PA that, in the experience of a sensuous group, the sense data somehow 'fuse' (*verschmelzen*) with each other. The term 'fusion' is chosen, Husserl explains, to stress that 'we grasp the quasi-qualitative moment of the whole intuition as something simple, not as a collectivum of contents and relations' (PA 204). In EU this term is still used in connection with the phenomenon of sensuous groups, though with a certain qualification. The sense data are 'fused,' Husserl says, but 'not in such a way that they flow over into one another, but in a kind of fusion at a distance (*Fernverschmelzung*), in which they can be made coincident with one another as being similar' (EU 76 [74]).

9 This 'turning toward' does not mean that the sense data themselves, as distinct from the objects they represent, are experienced as objects in their own right. It means only that they are interpreted or 'taken' as representatives of the actually intended objects. This notion of 'turning toward' sense data and 'interpreting' them as representatives of non-immanent objects was not introduced by Husserl until after the publication of PA. Thus it remains unclear in this early work whether or not the fusion of sense data corresponds to an 'objective' fusion of some kind (cf. LU II/1 234 [442]).

10 *Husserliana*, Vol. XI, p. 418.

11 Ibid.

12 Husserl seems to assume in PA that we either collect the items distinctly or settle for a merely 'symbolic' presentation of what we intend, a presentation 'through signs' (cf. PA 193). At one point Husserl writes: 'It is indubitable that the concrete multitude presentation is here not an authentic one, and that the subsumption under the general concept of multitude, which is associated with the name 'group' (*Menge*), could only have resulted in a symbolic way' (PA 197). Part of the problem here seems to be Husserl's assumption that when we experience a sensuous group, we actually 'intend' a distinctly collected multitude, not the sensuous group itself.

13 On signs (*Anzeichen*) as indications of what is not itself given with evidence, cf. LU II/1 24 f. (269 f.).

14 The importance of the phenomenon of sensuous groups is obscured somewhat by the organization of PA. Instead of beginning his investigation of the origin of number with a discussion of sensuous groups, as we have done in the present study, Husserl introduces this topic only in Part Two, in connection with the phenomenon of 'symbolic presentation.' Perhaps Husserl's aim in selecting this arrangement was to make vivid the point that number is not itself a 'visible and tangible phenomenon,' as Mill, for example, had claimed numbers themselves are visible or tangible. It does seem nonetheless, as Husserl's own later writings suggest, that a categorial object such as number is best explained by starting with a lower level phenomenon such as sensuous groups.

15 Cf. EU 134 (120): '... a plurality, for example, a cluster of stars, a cluster of colored dots – can also, on the basis of a unified prominence and affection, become a unitary theme, and its particulars can be *explicated* as determinate parts' (emphasis added).

16 In PA Husserl seems to be making this point when he says that quasi-qualitative moments are 'associated' with the process of apprehending the members individually (PA 203).
17 LU II/1 227 ff. (436 ff.); EU 160 ff. (140 ff.); cf. Sokolowski, *Husserlian Meditations*, pp. 9 ff.
18 EU 154 f. (134 f.); LU II/1 234 f. (442).
19 Curiously, Husserl does not develop the distinction between the 'pieces' of a thing and what we call the 'items' of a sensuous group in LU. In this work he takes pieces as the paradigm case of independent parts.
20 This seems to be what Husserl means when he writes in PA that 'it is untrue that we can *never* be occupied with more than a *single* content at one and the same moment.' When we distinctly retain previous acts of apprehension in the present moment of consciousness, we are in a sense 'occupied' with each of them simultaneously. It should be noted however that Husserl does not in his early works always observe the important distinction between the 'single glance' which takes in a sensuous group without articulation and the single moment in which previously executed articulating acts are distinctly retained (cf. PA 307).
21 See above, p. 32f.
22 In PA Husserl claims that multitude is actually *prior* to number. He claims that 'the concept of indeterminate multitude represents a considerably deeper level of concept formation' than does that of determinate multitude or number; it is a level already achieved by children and primitive peoples (PA 83). This might well be questioned. Sensuous groups, which are also of course indeterminate in respect to 'how many,' undoubtedly belong to a 'deeper level' than number; but multitudes in Husserl's sense are not sensuous groups. It seems more plausible to say that multitude and number both belong to the 'higher level' on which categorial activity comes into play. Multitude *is* distinguishable from number, however, and likewise the explicating act from the comparing act. It is this distinctness, not any priority of the one over the other, that justifies the separate treatment.
23 PA 336-338; see also above, pp. 39ff.
24 PA 81 f.; but cf. PA 83.
25 Torretti seems to miss this important point when he comments that 'para Husserl ... el núnmero de un conjunto es una caracteristica discernible en él sin necessidad de estudiar sus relaciones con otros conjuntos' ('La filosofia de la aritmetica de Husserl,' p. 192).
26 The German word *Gleichheit* can mean either 'likeness' or 'equality'; it is therefore necessary to determine from the context which is meant in any given case. Although Husserl uses *Gleichheit* in both senses, it is clear that he keeps these senses distinct. Indeed, in citing a passage from another author in which multitudes are spoken of as '*gleich*,' he suggests that '*gleichviel*' or '*gleichzahlig*' (which could be translated 'equally many' and 'equally numerous,' respectively) would better express the author's meaning (PA 98).
27 In PA Husserl describes comparing as a 'mental act of higher order,' i.e. as what he would later call a 'categorial act' (PA 92). In fact, he characterizes it as an act of the *third* order (PA 92 f.). (The discussion of this point is somewhat confusing, since he first characterizes the comparing act as an act of *second* order, but then goes on to correct himself.) Husserl's thought is apparently that 'first-order' acts make the sensuous objects that are to-be-counted present, while a 'second-order' act distinctly collects these objects and generates a multitude. The 'third-order' act then compares the resulting multitude with other multitudes. The comparing act is of the third order, in other words, because it is a categorial act that is founded in turn on another categorial act, not directly on straightforward sensuous acts. In this passage Husserl comes close to acknowledging explicitly what we may refer to as the three element of authentic counting: sensuous groups, explication and comparison.
28 In the case where a group is compared to itself at a different time, this 'likeness' actually becomes 'sameness,' of course.
29 Elsewhere in PA, Husserl uses the phrase 'in a single glance' (*in einem Blicke*), or some related locution, to characterize the presence of sensuous groups (PA 28 n., 196, 199, 205).

30 PA 403. On the dating of this manuscript, see the 'Textkritische Anmerkungen,' PA 532. The editor of the volume seems to err when he suggests that the reference in the passage cited is to Chapter VI of PA. Husserl is referring rather to Chapter VII (entitled 'Die Zahlendefinitionen durch Aequivalenz'), in particular, it seems, to the passage on PA 116 f.

CHAPTER IV

NUMBERS AS IDENTITIES IN PRESENCE AND ABSENCE

1. INTENDING NUMBERS IN THEIR ABSENCE

In the preceding chapter we were concerned with the original presence of number. Following Husserl's lead, we attempted to describe the 'authentic counting' in which numbers themselves are given in an originally intuitive manner. As we saw, numbers are not given in sense perception alone, but rather in a complex categorial activity that is founded on sense perception. Numbers are presented authentically when we take stock of the items of a sensuously presented group while comparing the resulting multitude with other multitudes. Now such descriptions of the original presence of number are an essential component of any study of the constitution of number. By themselves, however, they are not sufficient. For numbers have a way of being 'there' for us even when we do not have them as authentic presences. We can intend numbers in their absence as well as in their presence. In the current chapter, therefore, we shall study the absence of number. We shall see that the absence of number can take several distinguishable forms. We shall also see that our ability to intend numbers in their absence, however easily it may be overlooked, is of the greatest significance from a phenomenological point of view. Indeed, it seems that we cannot fully explain what we mean by 'number,' or clarify what numbers truly are, unless we take account of the fact that numbers are available to us as identities in presence and absence.

The presence generated through authentic counting is an achievement. It is this presence that first allows us to distinguish between sensuous groups and multitudes, and likewise between the different forms of multitude. But the ability to name and identify what is generated in authentic counting would seem to be yet a further achievement. It is conceivable that we could perform the categorial acts described above, that we could take stock of sensous groups and compare the resulting multitudes, without attaining that distance from what was thus presented that is an essential component of all naming.[1] We might explicate and compare without grasping 'the number' as something that somehow belonged to the multitude itself, as distinct from our experience of it.[2] To explicate and compare a group of four is one achievement; to see that the group itself is four in number is yet a further achievement.

See notes at the end of this chapter.

With very small groups, the ability to take the group as a number and the ability to name the specific number which pertains to it are scarcely separable from each other. When we can intend the group as a determinate 'how many,' as a multitude which stands in relations of equality and inequality to other multitudes, we can also state just how many there are: either two, or three, or four, etc. But with larger groups this is not necessarily the case. It is conceivable that we could intend such groups as identifiable numbers without actually being able to name the specific number which pertains. As Husserl puts this point: 'In a symbolic sense we can ... say of any group whatsoever that it amounts to a determinate number, even before we have formed this number itself – and indeed, even when we are not in a position actually to form this number' (PA 222). All naming of numbers involves achievement of distance on the part of the namer from that which is named; but the naming of large numbers involves skill and method as well.[3]

Where numbers can be named, they can also be intended in their absence. If the number of apples in the bowl, for example, is not authentically present to me, I can nonetheless refer to or 'mean' that number. I can still intend the group of apples as a fully determinate multitude. I might actually refer to 'the four apples in the bowl' without having this number genuinely present to me. Or I might simply ask 'how many' apples there are. Even in this latter case I intend the group as a number, though as a number which I am not yet able to name with full specificity. In both cases I am intending number in its absence. But there are several senses in which we may speak of the absence of number. Let us attempt to distinguish these.

One type of absence is bound up with the 'founded' character of the act of counting. In order for us to count authentically, there must be something there for us that is to-be-counted. A 'many' must be given in order for us to make the original determination of 'how many' (PA 15). The categorial act of counting, in other words, is inevitably founded on a lower level act of sensuous intuition. Husserl does not make this point in PA as clearly as might be desired, for in this work he takes his bearings from the arbitrary listing of items, not from what we have termed 'stock-taking explication.'[4] On the basis of Husserl's later writings, however, we have found it necessary to return to the passive givenness which authentic counting presupposes. From this point of view, it would seem that there can be no fully intuitive, authentic counting – that is, no original presence of number – where there is no group given as such in passive, sensuous fashion. If we use number names in the utter absence of sensuous groups, we use those names emptily. Of course, sensuous intuition can take several forms. Sense perception is the most obvious and most basic; but memory and even fantasy are relatively 'intuitive' modes of intentionality, in contrast to the most extreme form of empty consciousness.[5] It is possible for us to remember a group that was perceived earlier. And even if the items of the group were not counted as the group was being perceived, it is sometimes possible for us to count them as they appear in memory. The same thing is possible, at least within limits, for groups that have never been perceived at all, but only fantasied. Sensuous givenness can take many forms, but without some sensuous foundation, the original and authentic act of counting cannot be performed; number cannot become an authentic presence.[6]

There is a second form of absence that is closely related to the one just mentioned. This is a distinctive form of absence that arises when an empty number intention is 'frustrated' rather than 'fulfilled.'[7] Suppose I emptily intend the number of apples in the bowl. I name the number – either by saying that there are 'four,' or at least by asking just how 'how many' there are – even though the apples themselves are not sensuously present. In such a case I intend a number, a determinate multitude, in its absence. Now suppose I seek to bring this intention to fulfilment, to confirm that there are in fact four, or perhaps to determine the specific number for the first time. But instead of finding that there are four, I find that there is only 'one' or even 'none.' There is not a 'number' of apples after all; the apples do not in fact make a multitude, as I had anticipated. The 'one' or the 'none' that is now present to me is present precisely as a lack or 'privation' of number. It is not accurate to say of such a case simply that 'a thing' or 'nothing' is present. For what we find here is actually a peculiarly multi-layered or 'laminated' presence.[8] The previous number intention is somehow retained, but retained as something that was subsequently frustrated. Likewise, the one or the none itself is present precisely as something which had been expected to be 'many,' and indeed a determinate 'how many,' but which turned out not to be a many after all.

There is obviously something number-like about the presence of one and none in the situation described above. As Husserl points out in PA, the question as to 'how many' is as appropriately answered by 'one' or 'none' as it is by 'two,' 'three,' 'four,' etc. (PA 130 f.). An empty number intention can terminate, so to speak, in a presence of one or none, as well as in a presence of two, three, four, etc. But we should not for this reason speak indiscriminately of none and one as 'numbers' (*Anzahlen*).[9] 'None' and 'one' do indeed serve to answer the question as to 'how many.' But they are answers of a distinctive kind. As Husserl puts in PA, they are 'negative' answers (PA 130f.).[10] The presence of one or none, though number-like, is a laminated presence. Unlike the original presence of number, this presence presupposes that numbers have been named and intended in their absence. We encounter one or none only where an empty number intention has been frustrated rather than fulfilled. Had the one or the none been present to us all along, we would not have fount it to be number-like; we would not have asked 'how many.' However number-like none and one may be, we should not confuse them with numbers in the original sense. We should not confuse the presence of none or one – a presence is which number is somehow meant in absence – with the original presence of number itself.[11]

The two forms of absence discussed thus far arise from the fact that we can intend number in the absence of a sensuous foundation. But there are also other, more subtle forms of empty number intentionality. These latter arise not from the fact that there is no sensuous group that is to-be-counted, but rather from a failure to perform the categorial acts that are essential to the presence of number. The sensuous presence of a group does not by itself assure the presence of number;

authentic counting is required as well. We must actively explicate the group and compare the resulting multitude with other multitudes if number is to become an authentic presence. As Husserl warns in LU, 'we must ... guard against confusing the *straightforward percepts of sensuously unified groups (Mengen), series, swarms,* etc., with the *conjunctive percepts* in which alone the consciousness of multitude (*Vielheitsbewusstsein*) is itself properly constituted' (LU II/2 161 [799]). We intend a number in its absence whenever we intend or refer to a determinate multitude of items, but do so without actually counting the items in a fully energetic manner. We can of course fail *both* to explicate the sensuous group before us *and* to compare it with other multitudes. In this case we have a failure to perform either of the categorial activities that are needed to make number present. But it seems that we can also explicate without comparing and even, in a sense, compare without explicating. These latter cases call for further discussion.

It is possible for us to compare a given sensuous group to other groups without actually explicating the given group in full distinctness, without actually taking stock of the group. This could be said to occur whenever we merely associate a number name with a sensuous feature of the group in question.[12] We might, for example, survey the dots of a given die surface 'in a single glance,' and then come immediately to an estimate of the number of dots. For, as Husserl says, 'every die surface possesses a characteristic and fixed point-configuration which enters an association with the number name ...' (PA 252). Certain configurations are vaguely associated with the name 'three,' certain others with the name 'four,' and so on. When we are presented with one of these configuration, we utter the associated number name. In such cases we are obviously registering certain differences (for example, the difference between a triangular and a rectangular configuration of dots). In a sense, then, we are comparing the given group to other groups. But the relations we register are simply those of like and unlike, not those of equality and inequality – not, in other words, the relations that are essential to the authentic presence of number. When we associate a number name with a sensuous feature of a group, even if the number name is 'correct,' we inevitably intend the number itself in its absence.

Just as we can compare without explicating, so too, it seems, we can explicate without comparing. We can take stock of the items of a sensuously presented group in fully distinct fashion, but do so without actually comparing what is thus generated to other multitudes. When we do this, we make the group in question authentically present as an aggregate, but not as a number. We have a *multitude* present to us, but not a determinate multitude or *number*. In this case also, then, we intend the group's number in its absence if we intend it at all.[13]

Still another form of absence arises in those cases where we begin an authentic counting process, but do not carry it through to completion. Here we actually explicate the group in part, but nonetheless 'we do not achieve everything we could in the way of authentic activity' (PA 196). 'We omit the actual apprehension of all individual members, execute only a very few steps ... and thus are satisfied with a much more inauthentic subsumption under the concept of multitude than otherwise' (PA 196). In such cases, as Husserl puts it, we 'stop with the familiar 'etc." (PA 198). We can still intend the number pertaining to the group in question, of course. But where we do so, we intend it in its absence.[13]

We see then that there are at least six senses in which we may speak of the absence of number. We have, first of all, the case in which we intend a number, a determinate multitude, in the absence of the sensuous group in question. As a second form of absence closely related to this, we have the absence of number that is at work in the presence of what is only 'one' or 'none.' The third case is that which comes about when we have a sensuous group present to us, but fail to count it authentically; we neither explicate it nor compare it. The fourth and fifth forms of absence arise when we either compare sensuous groups without explicating them or explicate them without comparing them. In each case we perform one of the two categorial acts that are needed to make number present, but not both. A final form of absence arises when we count authentically, but do not carry our counting through to its completion.

It is extremely important to take account of the easily overlooked fact that numbers can be named and intended in their absence. As we shall see later in this chapter, intuitively empty number intentions are at work whenever we calculate with number signs and even when we 'count' the items of a group in symbolic, mechanical fashion. But there is also another reason why empty number intentions are important. According to Husserl, it is only with reference to the interplay of empty and filled intentions that the problem of the 'unity' of numbers can be adequately resolved. It is to this latter issue that we turn in the following sections.

2. THE UNITY OF NUMBER

In Chapter III we saw that numbers are authentically presented only through the performance of certain categorial activities. The presence of a sensuous group is not as such a presence of number. It is only when we actively explicate a sensuous group and compare the multitude it makes with other multitudes that number as such comes to be authentically present. Now these activities of explicating and comparing generate a new kind of presence, an explicitly 'plural' presence in which each of the items contained in the group is given in its own right. The group which had been present as a sensuous unity is now presented with full distinctness as a 'many.' But the explicit 'many-ness' of what is present in this way leads us to an important and difficult problem. A number is not just a 'many'; it is also somehow a 'one.' A number is somehow a unitary whole. But how can this unity arise in the midst of many-ness? How is it that a many can be one? This is an ancient philosophical question,[15] but Husserl claims it can be answered through careful descriptive analyses. As we shall see, Husserl's final, distinctively phenomenological solution to the problem of the unity of number depends on a proper appreciation of the interplay of presence and absence.

Husserl had recognized the unity of number as a problem already in his earliest works on the philosophy of arithmetic. As he points out in PA, there is 'an essential difference between multitude simply speaking, i.e. multitude thought as multitude, and multitude thought as unity' (PA 156). Even the most arbitrary aggregate (e.g. 'a feeling, an angel, the moon and Italy') is in some sense a unitary whole. 'The presentation of the aggregate is a *unity* in which the presentations of the individual objects are contained as partial presentations' (PA 20). In any aggregate or number, there is something which is somehow 'there' in addition to the items collected or counted. 'However easily one overlooks it, something is indeed there beyond the individual contents, something which can be noted and which is necessarily on hand in all cases in which we speak of an aggregate or multitude: the *connection* of the individual elements to the whole' (PA 18). Now this could of course be said of other wholes as well. There is a certain connection characteristic of the whole formed by a thing together with its features, and likewise a connection characteristic of that formed by a thing together with its pieces or 'physical parts' (PA 19f., 68 f.). But there is a fundamental difference, Husserl claims, between the connection found in the latter wholes and the connection found in collective wholes. In non-collective wholes, the connection is 'given along with [the parts] as a moment of the same presentational content' (PA 69). In such wholes 'the union is *intuitively noticeable* in the presentational content' (PA 72). That it to say, as the 'parts' come to be given, their connection to the whole comes to be given as well. But the 'collective connection,' as Husserl calls it in PA, is not given in this same manner. The unity of number is not a sensuous unity. A sensuous group as such might indeed be said to have a sensuous unity (cf. PA 195, LU II/2 161 [799]). But the characteristic form of unity with which we are now concerned is not of this type. The unity of a number or an aggregate is a supervening unity, not an underlying unity; it is a unity which presupposes many-ness, not a unity out of which many-ness first arises.

How is this supervening unity of a number or a multitude to be explained? Although Husserl had already recognized this as a serious philosophical problem at the time he wrote PA, he was not yet able to provide a satisfactory solution.

According to the Husserl of PA, the unity of a multitude or number is nothing other than the 'mental act' in which the items are collected (PA 73). The synthesizing act, he claims, is itself the unity of the collective whole. This act is a part of what we refer to when we speak of multitudes or numbers as unities. 'The collective union ... has its subsistence only in certain mental acts which encompass and unify the contents' (PA 73). A unitary aggregate arises when 'a unitary interest and, at the same time, in and with this, as unitary noticing, singles out various contents and embraces them' (PA 74). The unity of a multitude or number, in other words, is identical with the unity of the collecting act which engenders it.

According to the early Husserl's not altogether implausible account, the unifying act in question is to be considered an act of 'higher order.' It is not to be confused with the 'single glance' in which a sensuous group is given in straightforward, inarticulate fashion. Unlike an act in which we perceive a group sensuously, the act which provides the unity for a collective whole presupposes that, in addition to performing the unifying act itself, we also 'take hold of each of the members ... as something noted in itself' (PA 192).

> For the apprehension of each of the collected contents, a separate mental act is required; their synthesis then demands a new act which manifestly includes those articulating acts in itself, an act which therefore forms an act of *second order* (PA 74).

This higher-order act is thought of by Husserl as occurring simultaneously with the lower order acts on which it is based. Higher-order acts in general are, he says, 'directed at other mental acts in turn, and toward the primary contents only through mediation of these latter' (PA 92). The lower-order, articulating acts are retained in consciousness, so that they are all present together in a single moment of consciousness. The higher-order act then embraces and unifies these, thus providing the unity of the collective whole as such.

The unsatisfactory character of this account of the unity of numbers and multitudes has been alluded to frequently by commentators on PA. Indeed, Husserl himself later criticizied it, as we shall see below. But it is important to be precise about the defects of the account in question. The problem is *not* that Husserl reduces multitudes and numbers as such to 'mental powers' or 'mental entities.'[16] What is at issue here is only the *unity* of multitudes and numbers. The unity of the items with each other, or their 'collective connection' with the whole, is only one aspect of the collective whole as such.[17] The difficulty is not that Husserl reduces multitudes and numbers to mental entities, but rather that he construes them as curious halfbreeds, as wholes consisting partly of objects and partly of mental acts. He suggests in effect that when we speak of 'the four apples in the bowl,' we are referring not only to the apples themselves, but also to a mental act which collects them. To have an authentic presentation corresponding to the phrase in question, we would therefore have to experience the apples in distinct, articulate fashion, then turn inward and reflect on a mental act. The collective connection itself is a part of the number we refer to, yet this connection is given only in 'a presentation which reflects on the relating act' (PA 69). It follows that if we merely take stock of a sensuous group and relate the resulting multitude to other multitudes – remaining in a straightforward, object-directed attitude all the while – we do not achieve an authentic presentation of the number in question.

Husserl's basic error here seems to be a mistaken assessment of the achievement of reflective, act-directed thinking. It may well be that it is necessary to think about acts if we are to clarify philosophically what we mean in our talk of numbers and multitudes. But this is by no means equivalent to saying that the acts are actually referred to in such talk. We do not actually name or intend acts when we speak of multitude or number, though of course we must perform them if what we intend is to be made authentically present. It is only when we reflect philosophically that the acts performed in authentic counting become thematic in their own rights. Because the early Husserl failed to appreciate this fully,[18] he was led to regard reflection on acts as an intrinsic part of authentic counting and collecting themselves, and thus to identify the acts thereby disclosed with the unity of what is counted or collected.

In LU Husserl moved forward to a more adequate, though still not fully satisfactory view of the unity of number. In this work he no longer considers the unity of a multitude or number to be identical with the unity of the collecting act. The 'collective connection,' he has come to see, is something that is itself somehow objective. It is not objective in the manner of sensuous objects, but neither it it to be confused with the unity of the collecting act itself. A number is not a half-breed consisting partly of objects and partly of a mental act which unites the objects. When we speak of numbers, we do not covertly refer to mental acts – even though Husserl would still of course contend that reflection on acts is essential if we are to clarify what we mean by number. Let us examine the doctrine presented in LU more closely.

From the point of view of LU, the crucial observation is that the collecting acts is an act in which certain meaning intentions come to be *fulfilled*. What is actually given in the collecting act is something that could also be meant in an empty or signitive manner. There is a certain sameness, a certain identity, of what is merely intended in the one case and what is 'itself given' in the other.[19] It is because of this sameness or identity that the unity of number must be distinguished from the unity of the collecting itself. We see the theme of fulfilment clearly exemplified in the treatment in LU of the conjunctive particle 'and.' In PA Husserl had said that this expression serves to 'indicate' (*andeuten*) the collective connection. 'In and of itself it is without meaning; but where it binds two or more names, it indicates the collective connection of the designated contents' (PA 75; cf. 79 f.). The suggestion here seems to be that 'and' does not name anything at all, but merely commands a certain performance. In LU however this interpretation is abandoned in favor of another. Husserl now says that syncategorematic expressions such as 'and' *do* have meanings. They do name something which can be intended either in presence or in absence. To be sure, the presence of what 'and' names is not entirely similar to the presence of what is named by the terms it connects.[20] Nonetheless the meaning intention that is at work in the expression 'and' can somehow be brought to fulfilment: 'the acts in which ['collectiva' and 'disjunctiva'] are constituted as data are those which furnish a fulfilling intuition for the meanings of the conjunctions 'and' and 'or'' (LU II/2 160 [798]). In actually executing an act of collection, we 'bring to fulfilment ... a meaning of the *a and b*' (LU I 314 [508]). The authentic presentation of a collective whole, then, is to be understood as the presentation of something that could also be intended emptily or 'inauthentically.' The whole does not cease to be a whole when we no longer have it as an authentic presence; rather, it comes to be intended in its absence.

By viewing the problem of the unity of multitudes and numbers in the context of the broader issue of empty and filled intentions, the Husserl of LU is able to effect an essential improvement over his earlier treatment of this problem. Nonetheless the approach taken in LU also gives rise to a new problem. In LU Husserl contends that '*perception* and *object* are concepts which cohere most intimately together' (LU II/2 138 [78]). For everything that is objective, he holds, there must a correlative act of 'perception' in which the objectivity comes to be given. And since the unity of number is now acknowledged to be objective, there would have to be, it seems, a correlated form of perception. Obviously this will not be a form of 'sense perception,' for in an aggregate or number 'no material form of connection develops among [the collected objects] through the unitary intention' (LU II/1 282 f. [480]). There might of course be a sensuous unity which *also* obtains among the members; there might be 'an immediate bond of the representing sensuous contents themselves,' what Husserl would later call a passive synthesis of sense data. But this should not be mistaken for the unity of the 'collection' as such (LU II/2 174 [810]). The perception of a sensuous unity has 'no need of an articulated grasp and knowledge of individual items [and] does not therefore as yet possess the character of a genuine intuition of the collection as such' (LU II/2 161 [799]). The unity of a number or multitude, on the other hand, is a unity whose authentic presentation *does* require an articulated grasp of the individual items. It is not a sensuous, but rather a 'categorial' unity (cf. LU II/1 282 [480]). But if this categorial unity is truly objective, and if objectivity and perception are correlated as Husserl holds them to be in LU – then there must a distinctive kind of 'perception' in which the unity in question is given. There must be a form of 'categorial perception' or categorial intuition in which objects such as numbers and multitudes are given as unities.

In LU Husserl contends that we do in fact 'perceive' the categorial unity of a number or multitude. The structure of the intentional act correlated with this unity is, he claims, entirely analogous to that of acts of sense perception. An act of sense perception is distinguished from the corresponding empty intentions by the presence of sense-data which serve to 'represent' the object in question.[21] And there is, Husserl claims, something similar at work in the categorial perception of a multitude or number. What could it be that serves as a representative of the object in this case, where there can be no question of actual sense-data? According to Husserl, the 'mental bond' (*psychisches Band*) of the articulating acts to one another can be regarded as the representative in question: 'we believe that we may reduce the mental bond experienced in *actual* identification, collection, etc. ('actual' = authentic, intuitive) to a *universal common feature* ... yielding ... the representative content that pertains especially to the moment of categorial form' (LU II/2 173 [809]). This position is not at all easy to understand, but it is clear at least that Husserl does not in LU envision the 'mental bond' in question as actually *being* the moment of categorial unity, as he might have done in PA.

> A connection of acts is not, ... *pro tanto* a connection of objects. At best it can help to make such a connection apparent: it is not itself the connection that appears in it (LU II/2 172 [808]).

Just as sense-data are 'interpreted' to bring about an act of sense perception, so also it is the case the 'complete synthetic intuition arises ... in so far as the mental content which binds the underlying acts itself sustains *interpretation* as the objective unity of the founded objects. ...' (LU II/2 177 [812]).

But just what is this mental bond that is generated in an act of collecting? What is it that comes to be apprehended or 'interpreted' in the perception of a collective whole, in the way sense data are apprehended in a case of sense perception? This crucial question is never addressed in LU. Husserl never explains how authentic collecting can bring about a new mental content, a content which can then be interpreted as the representative of the collective whole in question.[22] It may therefore be said that Husserl's claims regarding the categorial perception of collective wholes remain unsubstantiated in LU. He does not fully resolve the issue of the unity of number in this work. LU does indeed represent a stride beyond the doctrine of PA, for here Husserl does at least make a distinction between the unity of the collecting act and the unity of the collective whole itself. But he is nonetheless unable to explain this latter unity in a fully satisfactory manner.[23]

It was apparently only after the publication of LU that Husserl advanced to what can be called his final position on the problem of the unity of number.[24] According to Husserl's final position, the unity of a categorial object is not something that is actually 'perceived.' The structure of the act correlated with this unity is not analogous to that of an act of sense perception. In sense perception, the sense data and the act that apprehends them are thought of as occurring simultaneously. But the act that consitutes the unity of a categorial object is, according to Husserl's final view, one which occurs only subsequently to the articulating acts on which it is founded. The plural consciousness as such is only preliminary, as it were, to a subsequently occurring 'backward glance' in which the collective whole emerges as a unity. In *Ideen* Husserl explains this as follows:

> The synthetically constituted collection thus becomes objective in the pre-eminent sense (i.e. becomes the object of a straightforward doxic thesis) in the backward reference of a straightforward thesis to a collection previously constituted in an original manner; that is, in the peculiar noetic attachment of a thesis to the synthesis. In other words: *plural consciousness has the essential possibility of being transformed into a singular consciousness* which takes from it the multitude as one object, as an individual ...[25]

We find a similar account in FTL:

> The *plural* makes its appearance in plural judging and, on being 'nominalized,' on being transformed into an object in the pre-eminent sense (the substrate, the 'object-about-which'), yields the *set* (*Menge*).[26]

And yet again, in EU:

> In order for the collective connection, originally sprung from the act of plural explication of *A* and *B*, to become a substrate – i.e. a true object, something identifiable – a *turning of regard* is first required. But this implies that, as long as we carry out a merely collective assemblage, we have, more than ever, only a *preconstituted object*, a 'plurality,' and only in retrospective apprehension, following the active constitution, do we have *as an object plurality as unity: as set* (EU 293 [246]).

We see then that the later Husserl has moved forward to a fundamentally new view of the supervening act which generates the unity of a multitude or number. The act correlated with this unity occurs *after* the articulate counting or collecting, not together with it. Authentic counting or collecting does not by itself constitute a unitary collective whole. The collection emerges as unity only when, having already performed an articulate categorial act, we come to intend what this act generates in 'nominal' or 'monothetic' fashion.

This view is different from those which Husserl had presented in PA and LU. In his early works, Husserl had seen a close relationship between the unity of the articulate, collecting act and the unity of the collective whole itself. In PA he had actually identified these. In LU he had at least claimed that a 'mental bond' which arises in the collecting act somehow founds a perception of the corresponding objective unity. But in his later works, the unity of the collection itself is no longer understood to be correlated with the unity of the articulate collecting act. We see this very clearly in the following passage from EU:

> The collective synthesis, the 'A and B and C,' is indeed the noetic unity of a consciousness, but not yet the unity of an object in the proper sense, that is, in the sense of a thematic object-substrate. Here A, then B, then C is thematic, but the collective is not yet thematic. The collecting consciousness contains several objects encompassed in unity, but not a unique object having several members (EU 294 [246]).

There is no objective unity, in other words, prior to the performance of the nominalizing act. Of course, we might anticipate performing this act even as we collect or count authentically. But the important point is that the object does not derive its unity from any unity that may be present in the articulate categorial act as such. It is only when we move beyond the categorial act and take the presence that it generates as something nameable and identifiable that we have the collection as a unity.[27]

Husserl's final solution to the problem of the unity of number can be reformulated in terms of presence and absence. According to this solution, we can say, a number is a unity precisely because it is an identity in presence and absence. What the subsequent nominal act intends it 'the same' as what was generated in the corresponding act of authentic counting and collecting. What is meant in the one act is identical with what is presented in the other. Where we actually perform the categorial act, we have the object as something intuitively, authentically present. Where we do not, we intend the object in its absence. It is precisely in light of this interplay of presence and absence that the unity of number becomes explicable. Authentic counting generates a 'many'; but this many is a 'one' only to the extent that it can be intended as something identically the same in presence and in absence. If a number could not be intended emptily, it would indeed be many, but it would not be one as well. The peculiar unity in the midst of multitude that is characteristic of number arises only through the interplay of presence and absence.

3. THE UNITY OF LARGE NUMBERS

In the previous section we described three stages in the development of Husserl's views on the unity of number. In the first two stages (represented by PA and LU, respectively), Husserl explains the unity of number with reference to the unity of the act of authentic counting. In the first stage (represented by PA), Husserl actually identifies the unity of number with the act of counting, while in the second (represented by LU), he claims merely that an objective unity is perceived in this act. It is only in a third stage (represented by *Ideen*, FTL and EU) that Husserl comes to a satisfactory view of the unity of number. Here he claims that this unity is something constituted in a 'nominalizing' act, an act which does not necessarily occur simultaneously with the acts on which it is based. Now in the present section we shall see that Husserl's final view of the unity of number allows him to deal with a problem which cannot be resolved on either of his two earlier approaches: the problem of the unity of large numbers. It is only when we come to see the unity of number as something constituted in a nominalizing act, or in the interplay of presence and absence, that we can explain how large numbers can be unities as well as small ones.

In PA Husserl distinguishes between 'authentic' number concepts and 'symbolic' number concepts. Small numbers like, two, three and four are said to be authentic concepts because they can be presented authentically. We may not actually achieve an authentic presentation of these small numbers in all cases (cf. PA 216); nonetheless the possibility of such a presentation is at least available. But Husserl claims that the situation is different with large numbers. Here we cannot achieve an authentic presentation, he contends, even if we attempt it. 'Only in especially favorable circumstances can we have authentic presentations of concrete multitudes of about a dozen elements ... Accordignly, twelve (or some smaller number close to it) is the final limit for the conception of the authentic number concepts' (PA 192; cf. 222, 224). All of the numbers greater than twelve, Husserl suggests, are in fact symbolic concepts which cannot possibly be presented authentically (cf. PA 7, 179, 222 ff.). They are numbers 'whose authentic presentation is denied us forever' (PA 223).

Why does Husserl make this sharp distinction between large and small numbers? Why does he say that large numbers are symbolic concepts that cannot possibly be given in authentic presentation? He certainly does not wish to deny that, where a large sensuous group is given, we can explicate it by running through its items one by one (cf. PA 214 f.). Nor, for that matter, does he deny that we could compare large groups and thereby form a judgment regarding their relations of equality or inequality.[28] The problem is rather that, where the group to be counted contains more than a dozen or so items, we cannot 'actually embrace each of its members in a single act together with all the others, as something noted in its own right (as it is intended)' (PA 192). The problem, in other words, is simply that the act of counting itself lacks unity in such cases. We cannot actually gather or collect all of the lower-order, articulating acts in one higher-order, supervening act. The reason for this is of course the limitation on our capacity for retention:

> We can no longer retain the successive apprehensions of the group's members in a single act. Only a small number of these remain at any given moment in the realm of collecting activity. As new members are apprehended and added on, others already singled out fall away; the acts in which they are presented become more and more blurred in the background of consciousness, and finally they disappear altogether (PA 198).

The lower-order acts cannot all be retained where the group in question is a large one, and thus they cannot be gathered in a single higher-order act. But since it is only in a higher-order act of this sort that the unity of number is achieved, according to the doctrine presented in PA, the unity of large numbers must always elude us. Large numbers must have a merely symbolic mode of being; as Husserl says, they must be mere symbolic concepts.

Now Husserl is undoubtedly correct in pointing out the limitations on our ability to retain the lower-order acts performed in authentic counting. When we run through the items of a large group one by one, the items apprehended earlier do in fact tend to lapse into obscurity as new items come forth into the foreground of consciousness. And if the unity of a large number were presented only in a higher-order act which gathers distinctly retained lower-order acts, then this unity would indeed always elude us. But despite its descriptive plausibility, the doctrine of large numbers presented in PA leads to certain paradoxes. If a number is a unity of some kind, and if the unity of a number is identical with the unity of the higher-order act; and if, finally, this act cannot be performed in the case of large numbers – then it is difficult to see how there could even *be* large numbers at all. Large numbers would seem to have no being whatsoever, if the act which is their unity cannot be performed.[29] Needless to say, Husserl does not actually draw this extreme conclusion in PA. He speaks rather of there being 'numbers in themselves' (*Zahlen an sich*) that are merely 'inaccessible' to us (PA 223, 260). But it is difficult to see how this contention could be sustained, given the doctrine of the unity of number adhered to in the work.[30]

In LU, where Husserl no longer identifies the unity of number with the higher-order unifying act, this most severe paradox is avoided. Husserl would still be forced to say that large numbers cannot be authentically presented, it seems; for where the articulating acts cannot be retained with distinctness, they cannot generate distinctive 'mental bonds' either. But he could maintain that there genuinely *are* large numbers nonetheless. There are large numbers, he could say, even though they can never be given intuitively; they exist, but only as absences to which there corresponds no fulfilling mode of presence. This view would not be as extreme as that suggested by the doctrine of PA, but it is paradoxical nonetheless.[31] For surely there is a significant difference between intending a large number in a straightforward, purely signitive act, and intending the same number in an articulate categorial act. If we take in a large sensuous group in a 'single glance,' or if we merely hear someone state the number of items in the group, it would indeed seem appropriate to say that the number is intended in its absence. But if we actually count the items for ourselves, if we actually run through the members of the group one by one and compare the resulting multitude with other multitudes, we generate a certain presence that was not there before. To be sure, this presence is not the same as that which is available to us where small groups are in question. In explicating a large group we cannot retain each of the apprehending acts with full distinctness, as we can when we explicate small groups. But this should not be taken to imply that large numbers have no presence of their own. The phenomenological differences between large and small numbers could be compared to those between three-dimensional things and two-dimensional surfaces. We cannot experience a three-dimensional thing 'all at once,' in the way we can experience a two-dimensional surface. No matter what perspective we adopt, there is always a 'back side' that is not actually given. But we do not for this reason say that three-dimensional things have no presence at all, or that they have a merely 'symbolic' mode of being. Material things have a mode of presence all their own; and if this mode of presence differs from others, we should not for that reason consider it a mode of absence. The same, we suggest, is true of large numbers: they are not presented in the manner of small numbers, but there is nonetheless a mode of authentic presentation that is fully appropriate to them. We make a large number present in the manner appropriate to large numbers when we count it authentically, when we explicate a large sensuous group and compare the resulting multitude to other multitudes.[32]

Husserl's later account of the unity of number allows us to see how this can be so. It allows us to see how even a large number can be a unity, despite the fact that it is not given in a single, instantaneous higher-order act. From Husserl's later point of view, it does not really matter whether the lower-order apprehending acts are all retained distinctly in a single moment of consciousness. They could be retained, but they would not need to be retained in order for the number to be constituted as a unity. The founded act which generates the unity of the number is a 'nominalizing' act, not a collecting or gathering act as Husserl had claimed in PA and LU. This nominalizing act is something subsequent to, not simultaneous with, the apprehending acts on which it is founded. It glances in any case only 'backwards' toward the presence generated in the stock-taking process. The limits on our retentional capacity are therefore inconsequential in regard to the question of the unity of large numbers. What is essential is that counting a large group makes the group present in a new way, and that this new presence can be taken as something nameable and identifiable in its own right.

From Husserl's later point of view, we suggest, the sharp distinction introduced in PA between 'authentic' and 'symbolic' number concepts loses its plausibility.[33] What is essential is the distinction between authentic and symbolic number *presentation*, or between empty and filled number *intentions*. But this is a distinction which applies to large and small numbers alike. There is a difference between the presence of a large number and the presence of a small number. And perhaps one could say that the presence of small numbers is somehow prior to or more fundamental than that of large numbers. Nonetheless intentions which have large numbers as their objects can find an appropriate mode of fulfilment. Large numbers are identities in presence and absence just as small numbers are. Their unity is, on principle, no different from the unity of small numbers.

4. SEDIMENTED NUMBER MEANINGS

In the first section of this chapter we saw that our ability to intend numbers in their absence is closely related to our ability to name them. Naming involves an achievement of distance from that which is named, an ability to recognize that what is named is not the same as our experience of it. The naming of numbers in particular involves a recognition that 'the number' made present through the activity of authentic counting is in fact a feature of the group or multitude that is counted, not of the activity through which it is counted. But our ability to name numbers and intend them in their absence leads to a further form of intentionality which we have not yet taken into account. In this form of intentionality, we cease to take number words as names altogether. We do make active use of number words, but the activity in which we use them is not that of authentic counting. Number words are severed from the context in which they acquire their original nominal meaning. We use the word 'four,' for example, but we do not activate the original meaning of the word; we do not allow what 'four' names to become present, nor do we even anticipate such a presence. In this new form of number intentionality – a form in which what Husserl calls 'sedimented' meanings are at work – number comes to be absent, as we shall see, in a new and quite significant manner.

Before discussing the way sedimented meanings can generate a new form of the absence of number, let us examine the phenomenon of sedimentation in more general terms. Husserl discusses this phenomenon in a late essay entitled 'The Origin of Geometry.'[34] In this essay, of course, the primary theme is the intentionality at work in our dealings with spatial figures. In fact, however, much of what Husserl says in the essay is of a quite general character and is therefore applicable beyond the geometrical sphere as such.

Husserl calls attention in the essay to 'the important function of written, documenting linguistic expression' (K 371 [360]). Writing, he claims, has an important ability to effect 'a transformation of the original mode of being of the meaning structure,' to alter the character of the meaning which it at first serves merely to express (K 371 [361]). Language is what first makes it possible to share with others the insights achieved in original, authentic thinking. 'In the contact of reciprocal linguistic understanding, the original production ... can be *actively* understood by the others' (K 371 [360]). Language can function in this way because it involves signs that are 'sensibly experienceable'; as Husserl says of written signs in particular, 'it is always possible that they be intersubjectively experienceable in common' (K 371 [361]). Written signs and spoken signs alike can function to 'awaken' the meanings they express in those who experience them. But Husserl regards it as important to examine this 'awakening' function carefully, especially as it is found in connection with written signs. The awakening which such signs generate, as he points out, is not in the first instance a full re-activation of the meanings they express. 'The awakening is something passive; the awakened signification is thus given passively, similarly to the way in which any other activity which has sunk into obscurity, once associatively awakened, emerges at first *passively* as a more or less clear memory' (K 371 [361]). Now we can, generally speaking, move from this passivity to a full re-activation of the original meaning itself. We can enjoy again what was meant originally, and even do so perhaps with the original self-evidence. But in our context what is important is the other possibility which Husserl sees:

> There is a distinction ... between passively understanding the expression and making it self-evident by reactivating its meaning. But there also exist possibilities of a kind of activity, a thinking in terms of things that have been taken up merely receptively, passively, which deals with significations only passively understood and taken over, without any of the evidence of original activity (K 372 [361]).

In this new kind of activity, we come to do things with words that had not been done before. We manipulate the words in new ways, allowing them to enter relationships with other linguistic signs which they could not have entered before. The words themselves come to take on new levels of meaning. But despite the new meanings, the original meanings, as Husserl says, are at work in 'sedimented' form (K 371 f. [361]).

Now the 'sedimentation' process described here is by no means confined to the sphere of geometrical thinking. It is also at work in our dealing with number. It is possible for us to appropriate number words and put them to work in new forms of thoughtful activity even without fully reactivating the original number meanings. This happens, it seems, whenever we 'count' or 'calculate' in the usual, everyday manners. It would not be possible for us to engage in these activities if we had always to re-activate the full, original meanings of the number words we use. But we also could not engage in them if the original number meanings were not somehow at work in the signs with which we deal. In everyday counting and calculating, we may say, the original number meanings are at work in sedimented form.

Husserl does not explicitly discuss the 'sedimentation' at work in counting and calculation in any of his published works. In PA, however, he does call attention to the peculiar character of these activities as we find them in everyday life:

> ... In practical tasks of counting given groups, as in tasks involving calculative derivation of numbers from numbers, one can arrive at the solution in a purely mechanical manner: one substitutes the names for the concepts and then, utilizing the system of names in a purely external procedure, one derives names from names, and finally there result names whose conceptual interpretations necessarily yield the sought-for result (PA 239).

These 'mechanical' and 'external' procedures are understood in contrast to the original thinking in which numbers are generated as genuine presences. Instead of using number words to name what is generated in authentic counting, we use them rather as tokens to be manipulated in accordance with certain established rules. The result is what may be called a new form of the absence of numbers. The original meanings are still somehow at work; we could not count or calculate at all if groups had not at some time been authentically explicated and compared. We are still therefore somehow related to numbers. They are still 'there' for us, though only at a distance, so to speak. But what is truly distinctive about the intentionality at work here is that we cease even to *anticipate* a presence of the numbers corresponding to number words. We use number words, and use them actively; but we do not take them as names of something that could be made present in authentic counting. We do not take the number words we use as names of absent numbers. In the passage cited above (from PA 239), Husserl describes counting and calculation as activities in which 'one substitutes the names for the concept.' Now this formulation requires refinement, as we shall see; but it does serve to call attention to the displacement of number meanings that is at work in both counting and calculation as we ordinarily perform them. It suggests very strongly that these activities are both instances of what Husserl himself would later call 'sedimented' thinking.

Sedimented meanings are at work in both counting and calculation, but they are at work in different ways is the two activities. 'Calculation' as we perform it in everyday life involves the use of written number signs.[35] It is, as Husserl says in PA, an 'essentially sensuous operation' (PA 257).[36] By 'calculation' we generally mean 'every symbolic derivation of numbers from numbers which rests mainly on a rule-governed operation with sensuous signs' (PA 258). In calculating one 'derives signs from signs on the basis of the number sign system according to fixed rules, and then takes the result as the designation of a certain concept, the concept that was sought' (PA 257). Prior to the actual manipulation of signs, there is a 'translation of the initial thoughts into signs'; afterwards there is a 'translation of the resulting signs into thoughts' (PA 258). During the actual calculation itself, however, neither the initial thoughts nor the subsequent thoughts are at work. We merely follow the rules of calculation, aithout actually activating the full meanings of the signs we use. This is not to say of course that no 'thoughts' are involved at all. We must think in order to follow the rules of calculation. Husserl's point is merely that the thinking at work in the calculative manipulation of number signs is of a different character from 'original' thinking in the realm of number, the thinking in which signs serve merely to express the truths we register about numbers themselves.

As we have seen, Husserl sometimes describes the thinking at work in calculation by saying that it involves a substitution of 'names' for 'concepts.' Calculation becomes possible, he says at one point, when, 'in the given complex of concepts and names, one abstracts from the former and retains only the latter' (PA 259). But this seems imprecise. In calculation we do not actually have to do with 'names' at all. As we manipulate the written signs, we do not take them as names of numbers. Rather, we take them merely as sensuous signs to be manipulated in a certain way, in accordance with the established rules of calculation. As Husserl himself suggests at one point in PA, we take them as 'quite arbitrary, meaningless signs (like markers in a game, for example)' (PA 237 f.).[37] It is only after the calculative task strictly speaking has been performed that we one again take a number sign as a name.

Not all forms of calculation involve a displacement of number meanings such as that just described. Calculation with an abacus, for example, does not involve the manipulation of signs that could also be taken as names of numbers (PA 273). Nonetheless it is a remarkable and highly significant fact that we *can* at least calculate with signs of this type. And such calculation would not be possible at all if we could not cease to anticipate a presence of the corresponding numbers even while actively using number words; it would not be possible if we could not allow the original number meanings to lapse into the obscurity of something at work only in sedimented form.[38]

Since everyday calculation involves the use of written number signs, it provides a rather straightforward example of the sedimentation described by Husserl in 'The Origin of Geometry.' But with everyday counting, the case may seem to be less clear. Here we do not ordinarily make use of written number signs; we confine ourselves to spoken number words. Counting, moreover, is unlike calculation in that it involves actually running through the items of a given group in order to determine 'how many' there are. Nonetheless a certain displacement of number meanings is at work in everyday counting, a displacement not unlike that at work in everyday calculation. Let us examine this issue more carefully.

When we count the items in a group in the ordinary manner, we run through the items one by one and utter a number word for each item that we find (the number words being taken, of course, in the established sequence). Husserl at one point describes this procedure by saying that 'we proceed in such a manner as to co-ordinate the number names in a mechanical way with the members of the group to be counted, and then regard the name we need last as that of the number being sought' (PA 177). But he immediately adds the following comment:

> Actually the names serve us at first as a series of signs fixed in memory but void of content; for during the counting their conceptual content does not at all come into consciousness. Only after completion of the process, in consideration of its genuine purpose, does the (authentic or symbolic) number concept enter consciousness as the meaning of the resulting number word (PA 177).

Here Husserl seems to acknowledge that, during the actual counting process itself, we do not in fact take number words as names. The original meanings of the number words we use are not activated; a presence of numbers themselves is not anticipated. At another place in PA, Husserl suggests an appropriate way of characterizing the interpretation of number words at work in everyday counting. Such counting, he says, involves a 'step-by-step labeling (*Signierung*) of the members of the group through the series of number names' (PA 239). When we count, we take number words not as names but as *labels*. These labels are distinctive in that they can also be taken as number names. And at the completion of the counting process, of course, the final label is in fact taken as a name of the number of items in the group being counted. But during the counting process itself, number words are not taken as genuine names. With regard to the sedimented meanings at work, we might indeed say that in themselves they 'are' names; but during the counting process, these number names are not used as names of numbers.

In PA Husserl suggests that everyday counting does not actually generate a presence of number. It is, he says, a 'symbolic' procedure which leads only to a symbolic presentation of number (PA 222 f., 239). But this seems to be a mistake. Everyday counting does indeed involve the use of number words whose original meanings have come to be sedimented. In this sense such counting could be described as 'mechanical' and 'external.' But the sedimentation involved does not imply that the number named upon completion of the process is intended in its absence. We saw in Chapter III that number is made present when we explicate a sensuously presented group and compare the resulting multitude with other multitudes. And this is precisely what we do when we count in the ordinary manner. What is distinctive about everyday counting is merely that the multitude to which the counted group is compared is in fact the familiar series of number words. In counting we in effect establish a one-to-one correspondence of the group being counted with another multitude, the artificial but ever-available multitude of requisite number labels.[39] In establishing that the number of items counted is 'equal' to the multitude of labels needed in running through it, we achieve a presence of the number itself.[40]

Number words are of course not the only artificial devices that can be used in counting (cf. PA 245 ff.). Not all counting, not even all artificial counting, involves a displacement of original number meanings. Nonetheless the fact that counting *can* take this form is as remarkable and significant as is the previously discussed fact that calculation can take the form of manipulating written number signs. Like calculation, counting would not be possible if we could not move back and forth between two different ways of taking number words. In calculation, we take written number signs as though they were markers in a game; in counting, we take spoken words as though they were labels. In both cases, however, we achieve our purpose only when, as a subsequent step, we take the resulting number word as a genuine number name. In calculation we take the resulting number sign not just as one more marker in the game, but as the name of the number that was sought. In counting, likewise, the final number word needed is not just a label for the last item, but also the name of the number pertaining to the group being counted. Neither counting nor calculation would be possible, in the forms in which we know them, if we could not allow the original nominal meanings of the words with which we deal to lapse into the obscurity of something that is at work only in sedimented form.

When we count or calculate in the ordinary manners, we do not actually intend the numbers which correspond to our labels or markers. We do not enjoy them as authentic presences, nor do we even anticipate an authentic presence of number. It is for this reason that we may speak of the absence which arises in connection with these activities as a fundamentally new form of absence, different in kind from those discussed in the first section of this chapter. This new form of absence is of great importance. As we shall see in Chapters V and VI of this study, it is at work in many of the achievements of what we know as higher mathematics. But there is at the same time a certain theoretical danger in the move from original thinking to sedimented thinking. The danger is that the original 'things themselves,' the objects which the words in question were originally used to name, come to be so thoroughly obscured that they are actually 'forgotten.' We come to confound the significances that have accrued in the course of sedimented thinking with the original meanings which first make the sedimented activity possible.[41]

Husserl is in effect pointing out the possibility of such obscurity through sedimentation in his critique of certain 'nominalistic' theories of number in the Appendix to Part One of PA (PA 170 ff.). He mentions Helmholtz, for example, as an advocate of the view that 'numbers' are to be considered 'a series of arbitrarily chosen signs,' each of which is 'defined only by its position in the prescribed series (PA 171). On this view of number, *one* is merely the sign which comes first in the series, *two* the sign which follows one immediately, *three* the sign which follows two, and so on. Husserl criticizes this view in part on the ground that it involves a confusion of ordinal and cardinal numbers (PA 174). But what is of interest to us in the present context is his diagnosis of the causes of the error in question. The advocates of this view, he claims, 'have confined themselves to the external and blind process [of mechanical counting], mistaken its symbolic function and confused the signs and the things themselves (*Zeichen und Sache verwechselt*)' (PA 177). They have tried to explain what numbers themselves are without re-activating the sedimented meanings at work in everyday counting. Because each number word occupies a certain position in the series of number labels, they conclude that the very 'meaning' of a number word is simply its 'positional value' within that series (PA 174). They do not see, in other words, that the series of number labels is possible only because numbers themselves, the determinate multitudes made present in authentic counting, can be named and intended in their absence.[42]

Husserl also mentions a second factor which may have contributed to these nominalistic misinterpretations of the number concept: 'the opinion that only an interpretation of universal arithmetic as a coherent system of signs could eliminate all of the great difficulties associated with this remarkable science' (PA 178). The implication of this remark is plainly that, in Husserl's view, the difficulties regarding 'universal arithmetic' or analysis can be resolved in some other way. But just how did he think this could be done? We shall investigate this question in the chapters that follow.

NOTES

1 R. Sokolowski, *Presence and Absence. A Philosophical Investigation of Language and Being* (Bloomington, Ind.: Indiana University Press, 1978), pp. 27-31.

2 Husserl argues, against Frege and others, that numbers are predicated of aggregates or multitudes, not of 'concepts'; see above, p. 33 f.

3 Husserl studies the method or 'system' that we follow in forming number names in Chapter XII of PA; see especially PA 226 ff.

4 See above, pp. 51 ff.

5 LU II/2 56 ff. (712 ff.); Sokolowski, *Husserlian Meditations*, pp. 25, 35 f.

6 It is interesting to take note of one of Frege's criticisms of PA in this regard: 'Bei der kollektiven Verbindung sollen die Inhalte bloss zusammen gedacht oder vorgestellt werden, ohne daβ irgendeine Beziehung oder Verbinding zwischen ihnen vorgestellt werde ... Mir ist das nicht möglich. Ich kann mir nicht die Röte, den Mond und Napoleon zugleich vorstellen, ohne mir diese in einer Verbindung vorzustellen; z.B. die Röte von einem brennenden Dorfe, von der sich Napoleons Gestalt abhebt, vom Monde rechts beleuchtet' ('Rezension,' p. 185). Frege is right in suggesting that a sensuous foundation is essential to the presence of number. He errs however in denying the possibility of a categorial unity in multitudes and numbers.

7 On the terms 'frustration' and 'fulfilment,' see LU II/2 41 ff. (701 ff.).

8 Sokolowski uses the term 'laminated' in discussing identity synthesis in Husserl; see *Husserlian Meditations*, p. 28 f.

9 There are of course other senses of number (*Zahl*) in which 0 and 1 can be considered numbers, see below, p. 123.

10 Husserl argues against Frege's thesis that every possible answer to the question 'how many?' is a number. He claims that only 'positive' answers are names of numbers. 'One' and 'none,' he contends, are answers to the question 'how many?' in the way 'never' is an answer to 'when?' or 'nowhere' to 'where?' (PA 130 f.). Frege replies to this contention in his review of PA ('Rezension,' p. 189 f.). Surely Frege has a good point when he asks why there should be two 'negative' answers to 'how many?' rather than one, as in other cases. But his comment that one can use 0 and 1 just as one uses all the other numbers 'without special cautionary measures' is not relevant in the context. The issue is whether 0 and 1 are numbers in the sense of *Anzahlen*.

11 '*One* object is not a collectivum of objects; thus the assertion that 'one is there' is no assertion of number. And likewise *no* object is not a collectivum, and thus the assertion that 'none are there' is not an assertion of number' (PA 131). On this point, of course, Husserl is in fundamental agreement with the Greek philosophical and mathematical tradition; cf. Klein, *Greek Mathematical Thought*, pp. 49 ff.

12 Cf. PA 252 ff.; LU II/2 161 n. (799 n.).

13 Husserl does not call attention to this form of absence, but it is clearly implied by his thesis that, in order to form the numbers two, three, etc., 'we must classify multitudes of units' by making 'judgments about equality and inequality' (PA 95).

14 This form of absence is of great importance in generating the notion of an 'infinite aggregate'; cf. PA 218 ff., FTL 195 f. (188 f.).

15 J. Klein discusses several ancient attempts to answer this question in 'The Concepts of Numbers in Greek Mathematics and Philosophy' (unpublished paper delivered before the Philosophy Club of the University of Virginia, 6 March 1939). According to Klein, the *Pythagoreans* dealt with it by dividing numbers into classes such as Odd, Even, Even-Odd, Triangular, Square, etc. Each number has unity, the Pythagoreans held, because it can be assigned to one of these classes. Thus, for example, "six' is 'one,' or rather, any six things can be conceived as *one* group, namely 'six,' because the Form 'triangle,' which is one in itself, causes these six things to be one' (ibid., p. 6). On Klein's interpretation, *Plato* explained the unity of number in a different way. Plato recognized that 'the 'forms' do not explain the differences between

numbers under the same form,' and thus that 'Arithmetic cannot be sufficiently explained by itself' (ibid., p. 9). Plato therefore introduced the non-arithmetical notion of 'ideal' or 'eidetic' number (*arithmos eidetikos*) and explained the unity of other numbers with reference to it. As an example of this notion, Klein mentions that 'Being' is for Plato the eidetic number 'two,' since Being consists of Rest and Change together (ibid., p. 10). Apparently Plato introduced other eidetic numbers as well, though only up to ten (ibid., p. 11). Thus Plato's solution to the problem is that 'the arithmetical numbers have unity because they are images of the 'ideal numbers'' (ibid., p. 12). Klein goes on to observe that *Aristotle* rejected the problem of the unity of number altogether: 'Aristotle denies ... that there is any unity in a number of things. A number of things signifies many things and is not itself one at all ... The only possible 'unity' of a number is the unity of the unit which is subjected to the process of counting. The unity of six apples is 'apple'' (ibid., p. 13). The remark with which Klein concludes his essay is also interesting in the present context: 'However questionable the Platonic position with respect to numbers might be, it seems that Aristotle did not see the real problem of numbers – how many can be one. This problem yet awaits solution' (ibid., p. 13).

16 H. Pietersma claims that 'Husserl ultimately identified numbers with a mental power' ('Husserl and Frege,' p. 316). R. Torretti claims that, according to the Husserl of PA, 'los conceptos básicos de la aritmética – conjuncto, pluralidad, número – se presentan como conceptos de entidades mentales en su raiz' ('La filosofía de la aritmética de Husserl,' p. 190).

17 In BdZ Husserl states unambiguously that there is a 'difference between the concepts collective union and multitude': 'in the first case interest rests exclusively on the connection of the arbitrarily thought contents, in the latter case on the aggregate of these contents as a whole, i.e. the elements are intended to in their union' (PA 338; cf. 78).

18 In PA Husserl does mention in passing that 'only an exceptional, scientific interest' concerns itself with the collective connection as such (PA 75). But apparently he attaches no significance to this observation.

19 On Husserl's notion of 'itself-givenness,' see above, p. 36.

20 It is because of this difference in mode of presence that Husserl distinguishes in LU between 'independent' and 'non-independent' meanings (LU II/1 312 [506]). 'And' is said to have a non-independent meaning (LU II/1 305 [501]).

21 Husserl distinguishes in LU between the quality, the material and the representing contents of an act of perception (LU II/2 166 [804]). The corresponding empty intention is said to have the same quality and material, but to lack representing contents.

22 Husserl *does* attempt to explain how representatives come about in the categorial intuition of states-of-affairs (LU II/2 152 ff. [792 ff.]). But this account cannot easily be transferred to the problem of collective wholes. Moreover, it seems to involve some fundamental confusions even as it stands; cf. Tugendhat, *Wahrheitsbegriff*, pp. 119-123.

23 Husserl in effect admitted the inadequacy of the solution offered in LU when he later expressed disapproval of his earlier theory of categorial representation; see the Foreword written in 1920 to the second edition of the Sixth Investigation (LU II/2 v [663]).

24 This solution appears to be anticipated in LU at II/1 469 ff. (629 ff.). It is not at present clear however whether or to what extent this material represents a second edition addition.

25 *Ideen* 294 f. (310). We have modified Boyce-Gibson's translation of this passage rather extensively. – It should be noted that, in the second edition of LU, Husserl also makes reference to the 'single-rayed act' in which a collective whole emerges as a unity (LU II/1 482 [639 f.]).

26 FTL 113 (107); cf. 83 f. (79).

27 From this later point of view, we may say that the Husserl of LU had been right to stress that categorial acts can be fulfilled, but wrong to think of their fulfilment as a form of 'perception.' Tugendhat makes this point nicely: 'So läßt sich der wirklich und bleibende Gewinn der Lehre von der kategorialen Anschauung nur erkennen, wenn man den für Husserl eigentlich maßgebenden Begriff der Erfüllung von seiner Verbindung mit dem Anschauungsbegriff befreit' (*Wahrheitsbegriff*, p. 128).

28 See the discussion of one-to-one correspondence at PA 105 ff.

29 Cf. Torretti, 'La filosofía de la aritmética de Husserl,' p. 187: 'Que sentido puede tener la representación simbólica de un contenido o concepto, si por un lado, las operaciones que entran en la formación de su representación propia son parte integrante del mismo, y, por otro, es patente que tales operaciones no pueden llevarse a cabo?' See also Frege, 'Rezension,' p. 190 f.
30 Husserl does claim that 'the presented unity of the process' of running through the items 'can subsequently serve as the symbolic substitute presentation for the authentically pursued but unachievable unity of the actual collection' (PA 199). But this seems somewhat contrived; with large groups, the problem is precisely that we *cannot* form a distinct presentation of the process as a unitary whole.
31 The view in question would also seem to contradict Husserl's assertion in LU that 'object' and 'perception' are 'concepts which cohere most intimately together' (LU II/2 138 [781]).
32 The difference between small numbers and large number might also be compared to the difference between a chord, which is heard all at once, and a melody, which takes time to be heard. But here there is also an important difference: a melody is temporal in its very being, while with large numbers it is only presentation to which time is essential. The being of numbers themselves does not include time (cf. PA 32).
33 As G. Martin observes, 'Husserl selbst scheint dieser Unterscheidung später nicht mehr dieselbe Wichtigkeit beigemessen zu haben ...' (*Klassische Ontologie der Zahl*, p. 143). It may be that Husserl still attached importance to the distinction at the time he wrote LU; at any rate the examples given in this work of authentic number presentation invariably involve small numbers (cf. LU I 171, II/1 141 [180, 366]).
34 The essay was first published posthumously in 1939 under the title 'Vom Ursprung der Geometrie' and is reprinted as a '*Beilage*' in Vol. VI of the *Husserliana* series (K 365-386 [353-378]).
35 Even when we do 'mental calculation,' we first imagine the written characters before us and then imagine ourselves manipulating those characters in the accustomed manner.
36 Husserl distinguishes four different senses of 'calculation' at PA 256-258; see below, p. 108 n. 38. We are here concerned only with what is called the 'most common' one (PA 257).
37 Husserl later came to acknowledge that even the markers in a game are somehow 'Meaningful'; see LU II/1 68 ff. (304 ff.) and below, pp. 114 f.
38 It is important to note that in our form of calculation, the written number words are ideographic in character, not phonetic: when we calculate, we write '3,' not 'three.' The fact that we do this seem to have a certain phenomenological significance. As Sokolowski points out, 'in alphabetical languages the written words stand for the spoken words, so we must turn from the language to imagination to find pictures or instances or variants' of the object named (*Husserlian Meditations*, p. 85). An ideogram, on the other hand, is itself a pictorial instance of what it names (ibid.). Since an ideogram names itself, it becomes plausible, at least in the case of number names, to treat it as something that names *only* itself, and eventually as something that does not name at all. Historically speaking, this development seems to have occurred first in a culture in which both ideographic and phonetic writing systems were available; cf. Waismann, *Introduction to Mathematical Thinking*, pp. 50 ff.
39 Cf. Frege, 'Rezension,' p. 183.
40 Husserl's denial in PA that everyday counting generates an authentic presentation of number seems to be related to his views on the unity of number, which we have discussed in the previous section.
41 On the relation between writing and forgetfulness, see also Plato, *Phaedrus* 274E-275B.
42 Even Husserl seems confused at times by the sedimentation of number meanings. He says at one point in PA, for example, that numbers in the fundamental sense (*Anzahlen*) form a 'series' such that each is 'more by one than that which precedes it' (PA 176). But surely this is incorrect. The labels used in counting are related as 'before' and 'after,' but numbers themselves, or determinate multitudes, are related only as 'equal,' 'more' and 'fewer.' It makes no more sense to say that a multitude of five is 'before' a multitude of six than it does to say that yellow is before blue.

CHAPTER V

THE SENSE OF ARITHMETIC

1. IDEAL NUMBERS

In the preceding chapters we were concerned with some basic phenomenological problems relating to the constitution of number. Following Husserl's lead, we attempted to bring to light the fundamental intentional structures at work in our dealings with number. We sought to make clear how numbers are authentically presented, how they are given in the most original manner. We also stressed that numbers can be intended 'emptily,' and indeed that they can be somehow 'there' for us even when we do not actually intend them at all. Numbers, we suggested, are identities in certain modes of presence and absence. In Husserl's view, such investigations as these form an essential part of the philosophy of arithmetic. They are not, however, the whole of it. We saw in Chapter I of this study that Husserl's ultimate goal was to make a contribution that 'desideratum of centuries,' the 'true philosophy of the calculus.'[1] His aim was not merely to clarify the concept of number for its own sake, but rather to shed light on the puzzling problem of the nature of mathematical analysis. He sought to clarify the 'true sense' of this discipline and to give a theoretical justification for it and the broadened number concept that is essential to it. This twin project of clarification and justification is therefore the central theme of the remainder of this study.

In the present chapter we shall deal with some issues which are both of considerable interest in their own right and essential to a full understanding of Husserl's clarification of the sense of analysis. Analysis is not, on Husserl's view, a 'science of number.' Nonetheless it arises only as a modification of a more elementary kind of mathematical thinking in which numbers are in fact thematic. It is with the intentionality at work in this original arithmetic that we shall be concerned in this chapter.

Let us begin with a distinction which Husserl regards as of crucial importance. According the view presented in LU and later works, it is necessary to distinguish between concrete, individual numbers and the 'ideal' numbers which are their species or essences. It is necessary, for example, to distinguish between a concrete multitude of four things and the ideal Four itself (which we sometimes refer to simply as 'the number Four').

See notes at the end of this chapter.

> Ii is ... *evident* that when I say 'Four' in the generic sense, as, e.g., in the statement 'Four is a prime number relatively to seven,' I am meaning the species *Four*, I have *it* before my logical regard, and am passing judgment on it, and not on anything individual. I am not judging about any individual group of four things ... (LU II/1 140 [366]).

A concrete multitude of things – say, the four apples in this bowl – is a number that is in a way localized in space and time. Although not a 'real,' sensuously presentable object, this number is something to which there pertains a definite 'now and here.' But ideal numbers, as Husserl stresses, have no such 'now and here.' Every ideal number is a 'timeless unity' (LU II/1 141 [366]). We who intend them and perhaps have them intuitively present are located in space and time, but the ideal numbers themselves are not.

The distinction between concrete and ideal numbers is regarded by Husserl as extremely important. He has, as a result, been characterized by some as a 'Platonist' in the philosophy of mathematics.[2] There is no doubt some justification for this view, especially in light of the similarities which his concept of number bears to that of Plato himself.[3] But it would certainly be incorrect to take Husserl as a 'Platonist' in the sense of viewing ideal numbers as 'cut off from all links with the reflecting subject.'[4] Husserl's basic philosophical point runs in precisely the opposite direction. Despite the timelessness and placelessness of ideal numbers, it is nonetheless the case that their 'being' can be clarified only through phenomenological analysis. That is, we can explain what ideal numbers 'are' only by studying the intentional acts through which they are intended and 'constituted.' The mathematician himself, *qua* mathematician, does not of course need to think about these intentional acts. His task is that of registering essential truths about ideal numbers themselves. But in philosophy, where the objective is rather the clarification of the mathematician's achievement, reflection on the intentional acts in question is essential.

How are ideal numbers constituted? How are they given intuitively? In approaching these questions, it will be helpful for us to begin with a brief examination of Husserl's early views on this matter. It should be noted that in his early works on the philosophy of arithmetic, Husserl makes no mention of 'ideal numbers' as such.[5] This notion makes its first explicit appearance in his later work LU. The early Husserl does however distinguish between concrete numbers and what he calls 'number concepts.' According to the view presented in PA, what we refer to when we speak of 'the number Four,' for example, is a number concept (cf. PA 15, 82). Let us see how such concepts are held to arise.

We noted previously that Husserl actually presents two different versions of the origin of number concepts in PA.[6] According to the most thoroughly developed version, the number concepts emerge when we 'eliminate the indeterminateness' in the general concept of multitude (PA 81). The general concept of multitude in turn originates in a process of 'abstraction' in which we shift the focus of our attention from a concrete multitude to a certain feature it possesses. This process can take place, Husserl says, only where a concrete multitude is given intuitively. 'No concept can be thought without foundation in a concrete intuition'; and where the general concept of multitude is at issue, it is imperative that we have 'the intuition of some concrete multitude in consciousness, in which we abstract the general concept' (PA 79). But how does this process of abstraction take place? According to the general theory of abstraction presented in an 1887 manuscript,

> the abstractum emerges when a certain moment, side, property or feature is singled out from the total intuition by means of isolating attention. It is precisely this moment which all the concreta have in common, and it on account of this moment that we declare them to be objects of the concept.[7]

For example, if it were a certain species of the color red that were at issue, we would have the abstractum when we found it as a moment or feature of an intuitively given red thing (cf. PA 194). And the process is essentially the same when we abstract the concept of multitude. Here too we arrive at the general concept when we focus attention on a certain distinctive moment of a concretely given object of the concept. The only difference between abstracting the concept of multitude and abstracting a sensuous concept such as color is the peculiar character of the abstractum in the former case. What all multitudes have in common is simply the 'collective connection' (*kollektive Verbindung*) of the items to each other. It is this connection therefore which 'represents the abstractum on which the general concept of multitude or collective whole is based' (PA 78). We obtain the concept of multitude in general, then, by disregarding the particular nature of the collected contents while we 'hold fast to their connection' (PA 79).

> Individual contents that are in some way determinate are given in collective connection; as we now move to the general concept by way of abstraction, we do not attend to them as thus determined; rather, our main interest concentrates on their collective connection, while they themselves are considered and attended to only as some contents or other, each one only as *any something, any one* (PA 79).

But what is the collective connection? According to the doctrine presented in PA, we recall, it is nothing but the higher-order mental act of collecting, the act which unifies the multitude in question (PA 73 f., 77). Thus multitude, Husserl says, is one of several concepts which 'arise in reflection on mental acts' (PA 85). This does not of course mean that the concept of multitude is *identical* with the concept of the collecting act. As Husserl says explicitly, the collective connection does not make up the 'whole logical content' of the concept of multitude, but only its 'most essential component' (PA 78). The concept of multitude is that of 'something that *possesses* this abstract moment of collective connection.'[8]

Now according to the account presented most prominently in PA, once we have the general concept of multitude, no further 'abstraction' is necessary in order to generate the number concepts as such. We simply 'eliminate the indeterminateness' in the general concept of multitude, thus giving rise to a 'manifold of determinate concepts that are most sharply delineated from each other' (PA 81). It is to these concepts that we give the names two, three, four, etc. (PA 81).

But in addition to this account of the origin of number concepts, we also find in PA adumbrations of another, somewhat more plausible one.

> To derive the number concepts [Husserl states] it is not necessary to take the general and indeterminate concept of multitude as mediator. We come directly to these concepts from any concrete multitude whatever ... (PA 81).

Here the suggestion seems to be that number concepts are abstracted from concrete numbers just as the general concept of multitude is abstracted from concrete multitudes. It is not the case, in other words, that a single act of abstraction suffices for the entire realm of number concepts. Rather, it seems, we perform a separate act of abstraction for each distinct number concept. The number Three is abstracted from concrete multitudes of three, the number Four from concrete multitudes of four, and so on. Husserl does not describe this direct abstraction of the number concepts in detail. He says that its nature is 'fully clear,' given the foregoing description of the abstraction of the general concept of multitude (PA 81 f.). He does add the following brief description, however:

> Disregarding the particular features of the synthesized individual contents, one considers and retains each content only insofar as it is 'something' or 'one,' and thus, with reference to their collective connection, one acquires the multitude-form pertaining to the given multitude ... (PA 82).

But perhaps the nature of this abstraction is not quite as clear as Husserl suggests. For what, we might ask, is the 'abstractum' when we move from a concrete group of four to the concept of Four itself? It cannot be the collective connection alone, for this connection is common to all multitudes and in no way distinctive of multitudes of four. The abstractum here, it seems, would have to be the relations of more, fewer and equal which the given multitude bears to other multitudes. By focusing attention on this set of relations – and for the Husserl of PA, this would mean, presumably, focusing on the relating acts themselves – we arrive at the concept of Four, or at what we refer to when we speak of 'the number Four.'

In LU, Husserl implicitly rejects both of his earlier accounts of the origin of number concepts. Number concepts do not arise through a process of abstraction such as that described in PA, he now maintains – regardless whether we think of this process as leading to numbers directly or only indirectly, by way of the concept of multitude.[9] Husserl does of course continue to speak of 'abstraction' in LU; and his description of the abstraction of numbers in particular does have certain affinities with the second of the two accounts given in PA. Nonetheless there are some very important differences to be taken note of here.

First of all, the Husserl of LU clears up an important ambiguity that had been present in his earlier works. It is not entirely clear what it means to say that the number Four is a 'concept.' A concept is often understood to be the sense of meaning of a word, as distinct from the object which the word designates.[10] But surely the number Four is not the sense of the word 'Four.' It is rather what we refer to when we use the word 'Four' as a name. The number Four is what is given with evidence when we bring the meaning intention associated with the number word to fulfilment. For these reasons, the Husserl of LU speaks of the entities named by such phrases as 'the number Four' not as concepts but as 'objects' of a certain kind.[11] As we have seen previously, these numbers are not sensuous or material objects, but rather 'ideal' ones.

> Ideal objects ... exist genuinely. Evidently there is not merely a good sense in speaking of such objects (e.g. of the number 2 ...) and in conceiving them as sustaining predicates: we also have insight into certain categorial truths that relate to such ideal objects. If these truths hold, everything presupposed by their holding must have being. If I see that the predicate of my assertion pertains to the ideal object 4, then this object cannot be a mere fiction, a mere *façon de parler*, a mere nothing in reality (LU II/1 124 f. [352 f.]).

It is then primarily because we can grasp with insight the necessary truth of such statements as 'Four is an even number' – because we can register such truths in an authentic way – that Husserl speaks of the numbers referred to in them as 'ideal objects.'

But how exactly is such insight possible? How do ideal numbers come to be given intuitively? The most explicit answer Husserl gives to this question in LU is contained in the following passage:

> If we make clear to ourselves what the number Five truly is, if we conceive of it adequately, we shall first achieve an articulate, collective presentation of this or that set of five objects. In this act a collection is intuitively given in a certain formal articulation, and so as an instance of the number species in question. Looking at this intuited individual, we perform an 'abstraction,' i.e. we not only isolate the non-independent moment of collective form in what is before us, but we apprehend the Idea in it: the number Five as the species of the form swims into our conscious sphere of reference. What we are now meaning is not this individual instance, not the intuited object as a whole, not the form immanent in it, but still inseparable from it: what we mean is rather the *ideal form-species* ... (LU I 171 [180]).

A careful analysis of this passage will allow us to see just how the Husserl of LU had moved beyond his earlier views on the origin of number concepts.

In the first place, we note a basic similarity between this account and those presented in PA. In PA Husserl had written that 'no concept can be thought without foundation in a concrete intuition' (PA 79). This had been taken to mean that a concrete number or multitude must be given authentically, in fully articulate fashion, in order for us to perform the abstraction that leads to the number concepts. Likewise, in the passage quoted above from LU, Husserl says that we must begin with 'an articulate, collective presentation of this or that set of five objects,' if the number Five itself is to be given authentically. The straightforward givenness of a sensuous group does not suffice; we must perform the categorial activities which are needed to make a concrete multitude of five authentically present as such.[12] Husserl mentions only collecting activity here, but clearly the 'counting' activity in which we grasp the relations which the given multitude bears to other multitudes is essential as well. For, as Husserl says, the collection from which we derive the presentation of Five itself must be 'an instance of the number species in question.' And to have an instance of the number Five present to us, we must count; mere collecting makes the group present only as a multitude.[13]

Yet there is also an important difference between the account given in the preceding passage and the corresponding accounts in PA. This difference concerns the nature of the 'abstraction' which leads to ideal numbers. In PA, we recall, Husserl had said that abstraction consisted primarily in focusing attention on a distinctive feature of the object before us. To abstract the concept of multitude, for example, we focus attention on the 'collective connection' which binds the elements together – in other words, on the mental act which collects and unifies them. Now there are two reasons why the Husserl of LU no longer finds this account acceptable. In the first place, he no longer considers the unity of the multitude to be identical with a unifying act; the unity or connection of the items is itself an objective feature of the multitude.[14] Likewise, the relations of equality and inequality in which a multitude stands would also be considered to have an objective status. But in addition to this, what the Husserl of LU understands by 'abstraction' is quite different from what he had understood by this term a decade earlier. Mere 'isolating attention' by itself is not sufficient to make an ideal object present to us. Thus even if we were to concentrate on the relevant objective features of the concrete multitude before us (the collective unity of its items, the relations in which it stands to other multitudes), this alone would not yield an authentic presentation of the ideal number itself. As Husserl expresses it in the passage quoted above, 'we not only isolate the non-independent moment of collective form in what is before us, but we apprehend the Idea in it.' The number Five is not the multitude-form present in the concrete group of five, but rather the latter's 'ideal form-species.'

What Husserl refers to here as 'abstraction' is what he elsewhere in LU calls 'ideational abstraction' (cf. LU II/2 162 [800]). In ideational abstraction, it is not a moment of an individual, but rather ' its Idea, its Universal' that is brought to 'actual givenness.' What is authentically presented in such abstraction is something common to many individuals, not something that could be considered a moment of any one of them. Now Husserl distinguishes in LU between two basic types of ideational abstraction: 'we distinguish between *sensuous abstraction*, which yields *sensuous concepts*, ... and *purely categorial abstraction*, which yields *purely categorial concepts* (LU II/2 183 f. [818]). The difference between these two rests on the type of objects which serve as foundation in each case.[15] Sensuous concepts – for example, 'color, house, judgment, wish' – 'find their immediate basis in the data of sensuous intuition.' Categorial concepts or 'categories' on the other hand find their immediate basis 'in the data of categorial intuition, purely with regard to the categorial form of the whole categorially formed object.' Husserl does not specifically mention 'number' in his discussion of categorial abstraction, but since this concept is listed elsewhere in the work as one of the formal 'categories' (cf. LU I

244 [237]), clearly it is thought of by him as the result of an act of categorial abstraction.

Thus the act of 'abstraction' in which ideal numbers are given authentically is to be regarded as one of 'ideational' and in particular of 'categorial' abstraction. In other words, it is an act that is founded on categorial intuition, yet it does not consist simply in focussing attention on a moment of a presented categorial object. In what then *does* this act of abstraction consist? Can we describe more precisely how ideal numbers come to be given themselves?

Unfortunately, Husserl does not tell us in detail in LU just how the Idea of Five, for example, as distinct from the various individual fives that fall under it, comes to be presented authentically. He does not offer any analyses of the presence of ideal numbers more specific than those we have already examined. Nonetheless he does have something more to say about ideational abstraction in general, and what he says in this regard has obvious implications for his doctrine of categorial abstraction in particular, including that in which ideal numbers come to be self-given. In the Sixth Investigation, Husserl makes it clear that ideal species are given only when we recognize an 'identity' of sorts within a multiplicity of individuals falling under the species. Where only one individual is presented, we can indeed 'mean' the universal species or Idea. But the universal is actually *given* only 'through the repeated performance of such acts upon a basis of several individual intuitions' and through the 'overreaching act of identification which brings all such single acts of abstraction into one synthesis' (LU II/2 162 [800]). It is only in this way that the 'kind' (*Art*), as distinct from the many individuals *of* this kind, 'may *itself* come before us, and may come before us *as one and the same*.' Performance of the 'overreaching act of identification' makes it possible for us to grasp, for example, not only that 'this is a house,' but also 'what-it-is-to-be-a-house.' When Husserl speaks of our ability to 'behold' (*erschauen*) or 'intuit' essences, or when he says that 'the universal itself is given to us' (cf. LU 162 [800]), what he has in mind is an act of this sort, an act that grasps a unity in a multiplicity of presented objects.[16] If the multiplicity is not distinctly given, then the universal or species cannot be given either.

Applied to the problem that is of concern ot us here, this doctrine would mean that an ideal number is authentically present to us only when we grasp it as something that is identically the same in a multiplicity of particular numbers falling under it. We 'behold' the ideal Five, in other words, only when we become conscious of an identity present within a multiplicity of individual fives, of individual multitudes whose 'how many' is given precisely as five. We see explicitly that this particular five is 'the same' as that five and those other fives. Each of them is recognized as an instance of 'what-it-is-to-be-five.' When Husserl alludes in LU to the act of 'categorial abstraction' which makes an ideal number present, what he is referring to is nothing other than an act such as this in which we become conscious of the unity within a multiplicity of concrete numbers.[17]

The view presented in LU of the presence of ideal numbers is important in several respects. It shows that although there is a difference between empty and filled intentions of such numbers, the fulfilment in question can never take the form of a mere straightforward 'gazing.' Rather, ideal numbers are given only in a form of articulated consciousness in which we grasp the unity in a multiplicity of individual numbers of things. The presence of an ideal number is thus a categorial presence, and indeed in two different senses: first, in that the presence of *any* ideal species requires that the individuals of the manifold be actively synthesized, and second, in that in this case, the *founding individuals themselves* are given only through the categorial activities of counting and collecting. These points are quite important. They do not, however, amount to a complete account of the presence of ideal numbers. There is one crucial issue which Husserl fails to raise in LU, namely, that of the mode of generation of the manifold within which the ideal species is given as a unity. Husserl does not address this problem either as it concerns ideational abstraction in general, or as it concerns the abstraction of number in particular. To be sure, he does say at one point that 'our consciousness of the universal has as satisfactory a basis in perception as it has in imagination' (LU II/2 163 [800 f.]). In other words, according to LU it is a matter of indifference whether the manifold that underlies an act of ideational abstraction is generated through perception or through fantasy. If we wish to make the essence of 'house' authentically present to us, it does not matter whether we take as the basis of such presence houses that are actually perceived or houses that are merely imagined. But it is precisely this point that the later Husserl rejects. According to the doctrine presented most extensively in *Experience and Judgment*, the manifold of individuals within which the species or *eidos* is presented as an identity *must* be generated through 'free variation' (EU 410 ff. [340 ff.]). That is, rather than merely taking a manifold over from perceptual consciousness, we must actively construct a manifold of individuals through the free play of fantasy (cf. *Ideen* 161 f. [182]). We must take one individual as a 'model,' then vary the features of this model in as many ways as possible, up to the point where the variants cease to present themselves as 'similar' to the original.

> It then becomes evident that a unity runs through this multiplicity of successive figures, that in such free variations of an original image, e.g. of a thing, an *invariant* is necessarily retained as the *necessary general form*, without which an object such as this thing, as an example of its kind, would not be thinkable at all. While what differentiates the variants remains indifferent to us, this form stands out in the practice of voluntary variation, and as an absolutely identical content, an invariable *what*, according to which all the variants coincide: *a general essence* (EU 411 [341]).

An essence or 'idea,' then, is still said to be given as a unity within a manifold of individuals, even by the later Husserl. What is new here is only the insistence on a specific mode of generation of the manifold, on the need to construct the manifold through the free play of fantasy.

The passage cited in the preceding paragraph seems to relate most directly to cases where the essence in question is one whose instances are sensuous things. But not all essences or Ideas are of this type, as we have seen. There are also 'categorial essences' whose instances are themselves categorial objects. It is with such essences – and especially with the ideal numbers that form one class of them – that we are primarily concerned in the present context. It is therefore interesting to see that the late Husserl applies his revised doctrine of essential intuition to what he had earlier called 'categorial abstraction' as well as to what he had earlier called 'sensuous abstraction.' Free variation is a necessary aspect of the presence of categorial essences, just as it is in that of 'sensuous' ones.

> We not only can vary things of experience and thereby attain concepts of things as essential generalities, but we also 'experience' sets which we have collected independently, real states of affairs, internal and external relations, whose seeing requires an activity which relates them, and so on. In this way we also obtain pure and general ideas (*Ideen*) of collections, of relations, and of every kind of state of affairs, in that, starting from the intuitive activities in which they attain givenness, *we constitute precisely for all such objectivities multiplicities of variation which bring out the necessary and general essence* (EU 346 [359f.], emphasis added).

Unfortunately, Husserl does not develop this topic further. Neither does he relate what is said here to the problem of number presentation in particular. But the implication of this passage is clearly that numbers in the ideal sense, like the other categories mentioned, are themselves given only as identities within freely constructed manifolds of individual numbers. The ideal Five is given as that which is the same in all imaginable concrete sets of Five, and so on.

At this point a question arises. In what sense can we speak of constructing a variation manifold of numbers through the free play of fantasy? Have we not said that numbers are given only through the *categorial* acts of counting and collecting, and that these are not reducible to acts of sensuous intuition? And if this is so, how can we be said to 'imagine' numbers? We suggest that this question should be answered as follows. It is true that we cannot imagine a number as number. We can however imagine a group of items, and we can also imagine ourselves collecting and counting these items. Even if we have only heard descriptions of Mount Rushmore, for example, we can imagine the faces engraved in stone there, and we can also imagine ourselves forming a distinct presentation of the number of those faces. For that matter, we can even imagine groups of items that have 'nothing to do with each other' – for example, a group consisting of 'a feeling, an angel, the moon and Italy' (PA 16). And we can imagine ourselves counting even such arbitrarily concocted groups as these. It is in this way – by imagining groups of items, then imagining ourselves counting and collecting the items – that we are able to construct variation manifolds for numbers. The ideal number Four emerges as something that is present as identically the same in the group of apples in the bowl, the imagined group of faces engraved in stone at Mount Rushmore, and the arbitrarily concocted group consisting of 'a feeling, an angel, the moon and Italy' – as well as in all other imaginable groups of four.

When we register truths about ideal numbers – that is, when we do not merely repeat statements about them we may have heard or read, but actually verify them for ourselves – the evidence we enjoy is of the sort described in the preceding pages. Of course, we do not ordinarily reflect upon the evidence underlying our statements about number, even when we do enjoy it. We think about the numbers themselves and their relationships to one another, not about evidence as such. Nonetheless the intentional structures disclosed in Husserl's phenomenological reflection are there and at work all along. Without them we could not achieve the truths about number that we do in fact achieve.

2. THE FORMAL CHARACTER OF THE CONCEPT OF NUMBER

One of the recurring themes in Husserl's philosophy of arithmetic is his insistence that the concept of number, according to its own proper sense, is purely 'formal' in character. As such, he claims, it belongs with other formal concepts or 'categories,' including those of multitude, object, something, state-of-affairs and relation. It is characteristic of such categories that they are not restricted in their application to objects of a certain genus, but rather 'include in a certain manner each and every content in themselves' (PA 84); they are 'independent of the particularity of any material of knowledge' (LU I 244 [237]); they possess 'a formal universality that, on principle, leaves out of consideration every material determination of objects' (FTL 81 f. [77]).[18] In regard to the concept of number in particular, what these formulations are intended to bring out is simply that any group of objects – however diverse they may be, however improbable their grouping – can be 'counted,' presented as a number.

> In fact, for the formation of concrete aggregates there are no restrictions whatsoever in regard to the individual contents to be included. Every presented object, whether physical or mental, abstract or concrete, whether given through sensation or imagination, can be united with any other, with arbitrarily many other objects, to form an aggregate; and accordingly it can also be counted ... Thus the nature of the individual contents is in no way relevant (PA 16).

As Husserl puts it more succinctly in another place, 'nothing can be thought that could not also be counted' (PA 117).

How does the formal concept of number arise? Husserl sometimes says that it and other formal concepts are 'derivative formations of anything whatever' (*Ableitungsgestalten des Etwas-überhaupt*).[19] Now this would seem to be a somewhat misleading formulation, insofar as it suggests that the concept of 'something' (*Etwas*) is somehow prior to the 'derivative' formal concepts themselves. Husserl actually asserts this priority in PA. 'When we count,' he writes in this work, 'we bring each of the concretely present things under this concept [viz. the concept of something]' (PA 84). In other words, the very act of accounting requires us to consider each of the things to be counted as mere 'somethings.'[20] But if the concept of something is already presupposed by the act of counting, how does it itself arise? It is certainly not by 'comparing' all conceivable objects with each other and searching for what they have in common that we come to view them as mere somethings. As Husserl himself notes, 'such a comparison would simply remain without result' (PA 80). We cannot arrive at the concept of something by moving up the ladder of species and genera, so to speak, until we reach the very highest genus.[21] How then do we arrive at it? Husserl claims in PA that 'the concept of something owes its origin to reflection on the mental act of presentation, as whose content every definite object is given' (PA 80). It is when we regard the objects presented to us as presented or presentable, in other words, that we come to view them as 'somethings.'

Now certainly Husserl's negative claim here – that the concept of something does *not* arise through a comparison of objects – is a plausible one. But his positive thesis – that the concept of something arises through reflection on the act of presentation – seems somewhat less plausible. It appears to involve a certain confusion of the achievement of straightforward, object-directed thought with the achievement of reflective, philosophical thought. Undoubtedly it is possible to reflect on acts of presentation and to view objects as what is presented in them; such reflection is essential to philosophy as Husserl conceives it. But it seems incorrect to say that it is this philosophical reflection which itself generates the concept of something, the concept that is at work in number and other formal concepts. It would be more appropriate, we suggest, to say that the concept of something is *correlative* to the formal concepts in which it is ingredient, not prior to them. We take objects as 'somethings' only when we take them as ingredient in categorial objects.[22] Prior to the constitution of categorial objects, we do not take objects as mere somethings.

A further point must also be brought out here. In PA, Husserl seems to suggest that the 'original' concept of number is already a purely formal one. He claims at any rate that the formal universality of number 'already excludes a class of views on the origin of number concepts,' namely those which take numbers to be derived in some way from the world of sense experience.[23] Now certainly Husserl is correct in holding that sense experience alone cannot make number present. But he also seems to hold that as soon as we can speak of 'the number Four,' for example, we find it to have a purely formal content, a content that can be expressed by the formula 'one, one, one and one' (cf. PA 81 f.). And this, we suggest, might well be disputed. It seems possible to contend that the concept of number goes through an intermediate stage before arriving at the unrestricted generality of a purely formal concept.[24] This intermediate stage would correspond to what Husserl himself in other contexts calls an 'empirical universal.' An empirical universal, as he explains in *Experience and Judgment*, is a universal given as a unity in a class generated through our actual experience of the world, rather than through the activity of free variation.[25] As such, it does not yet possess the pure generality of an *eidos*. Now despite Husserl's failure to mention the concept of number in this regard, it would nonetheless seem quite plausible to say that an empirically based number concept precedes the emergence of the purely formal number concept.[26] At the stage where number is a mere empirical universal, we could indeed refer to 'the number Four.' We could talk about something that individual multitudes of four have in common. But we would not yet have arrived at a consciousness of Four as a pure essence or *eidos*. We would not yet have arrived at the insight that 'four' is a purely formal concept. It is only when we freely and imaginatively vary a given instance of Four that we begin to see that the concept of Four applies to a much broader range of concrete multitudes than those which emerge through the articulation of perceived sensuous groups.

This is not to deny of course that the *content* of the number concept, when the latter attains its own proper sense, is purely formal. It is only to point out that this formality is an achievement which goes beyond and builds upon a prior achievement. We arrive at a purely formal concept of number only through the process of eidetic variation described in the preceding section of this study.

3. ARITHMETIC AS FORMAL ONTOLOGY

In the preceding sections of this chapter, we have discussed Husserl's notion of ideal numbers constituted through free variation, as well as his corresponding assertion the number is a purely formal concept. We shall now proceed to examine his conception of arithmetic as such, the science of number which precedes and conditions the emergence of mathematical analysis. As we shall see in the present section, the notion that numbers are ideal objects possessing an unrestricted formal universality is central to Husserl's conception of this discipline. It is only because we can register necessities about numbers, only because we can perform acts of eidetic variation on concretely presented numbers, that arithmetic is possible at all. Let us examine this issue more closely.

There are some sciences, on Husserl's view, which aim only to establish contingent truths, truths about states-of-affairs that could be other than they are. The truths established in such sciences – which Husserls calls 'sciences of fact' (*Tatsachenwissenschaften*) – are based ultimately on the evidence of our 'actual experience' of the world (*Ideen* 11 f. [46]). But on the other hand, there are also certain sciences which aim to establish necessary truths about objects, truths regarding states-of-affairs which could not possibly be other than they are. These necessary truths cannot be established by our actual experience of the world; they rest on the 'essential insight' which arises when we perform the imaginative variation described in the first section of this chapter (cf. *Ideen* 13 f. [48 f.]). Thus Husserl refers to this second group of sciences as 'sciences of essence,' as 'eiedetic sciences,' or simply as 'ontologies' (*Ideen* 21, 23 ff. [55, 56 ff.]). As an example of such a science, he mentions classical geometry: 'geometry is an ontological discipline relating to an essential phase of [material] thinghood, the spatial form' (*Ideen* 25 [58 f.]). Now within the general sphere of eidetic science or ontology, Husserl distinguishes between 'regional' and 'formal' ontology. A *regional* ontology is an eidetic science which is directed toward some specific 'material' region (*Ideen* 23 ff. [57 ff.]). Geometry constitutes an example of regional ontology, as do the *a priori* theories of time and motion (cf. *Ideen* 21 [55]). *Formal* ontology, on the other hand, is not restricted in this way to any one specific material region; it is concerned in a way with properties that belong to every possible object precisely as such. Formal ontology, Husserl says, is 'the eidetic science of object in general.'[27] By this he means that it studies particular objects solely insofar as they are ingredients in 'categorial objects' such as facts, relations, sets and numbers (*Ideen* 28 [61]). Now in a sense, of course, even formal ontology could be said to relate to certain determinate regions. Set-theory, for example, is a branch of formal ontology concerned with the essential features of sets (cf. *Ideen* 362 [378]). Likewise, number-theory or arithmetic is a branch concerned with the essential features of numbers.[28] But Husserl's point is that the basic concepts of these disciplines (set, number, etc.) are themselves formal concepts, concepts of objects in which 'anything whatever' may figure as an ingredient.

It is important to stress that formal ontology is concerned with *objects*, not with meanings or judgments. We can of course subject the latter to formal study also; this is precisely what we do in the discipline which Husserl terms 'formal apophantis' (FTL 76 ff. [72 ff]). Formal apophantics, or what is usually called 'logic,' is concerned primarily with relations of consistency and consequence between judgments (propositions). But the 'categories of meaning,' of which judgment is one, must be distinguished from the 'objective categories' strictly speaking.[29] In formal ontology we are concerned only with the latter. Here our aim is not merely to formulate judgments that cohere with one another, but rather to disclose necessary truths about the categorial objects in question.[30]

Now all of these distinctions must be born in mind if we are to appreciate Husserl's claim that arithmetic is a branch of formal ontology. Its aim is to establish essential, necessary truths pertaining to the sphere of number – truths which can only be achieved through eidetic variation over the concrete numbers constituted through counting and collecting. But just what sort of truths do we discover in arithmetic? What exactly does arithmetic disclose about numbers? For answers to these questions, we must consult Husserl's early writings on the philosophy of arithmetic.

The Husserl of PA does not of course state that arithmetic is a branch of 'formal ontology.' As we have already noted, in this early work he considers numbers themselves to be 'concepts,' not ideal objects.[31] Nonetheless this work does contain what is by far Husserl's most detailed discussion of the nature of arithmetic. Many aspects of this discussion remain relevant within the context of his mature position; to appropriate them, we need only make the adjustments that are demanded by his later view of numbers as ideal objects.

What is arithmetic? According to Husserl, it is somewhat imprecise to define arithmetic simply as 'the science of numbers.' For, as he puts it, 'the individual numbers considered by themselves give no occasion for a scientific treatment' (PA 256). What needs scientific study is rather the *relations* between numbers. The essential task of arithmetic is that of 'finding other numbers from given ones by means of certain familiar relations' (PA 256). Arithmetic is therefore best defined as 'the science of number relations' (PA 256).[32] Now the 'relations' Husserl has in mind here are obviously those of equality and inequality – the relations which are definitive for the entire number region, as we saw in a previous chapter of this study.[33] Thus the essential task of arithmetic, it seems, is that of clarifying the relations of equality and inequality which necessarily obtain between or among numbers.

Why do we need a science of number relations? One reason which Husserl points out is that the relations in question are sometimes presented only with obscurity. It is sometimes unclear whether two numbers are equal or unequal. This is not true in all cases, of course. Numbers would not be available to us at all if we could not form authentic presentations of the exact relations obtaining between at least some multitudes. If we perform the appropriate categorial acts, we can 'see' that three is necessarily more than two, that two and three together necessarily make five, and so on. But not all numbers can be presented authentically in the way these small numbers can be. With large groups, we cannot retain all the acts of individual apprehension that are performed as we run through the items. Large numbers and their relations, therefore, cannot be presented in the way small numbers and their relations can be.[34] It is precisely this circumstance, according to Husserl, which gives rise to arithmetic in the sense of a science of number relations.

> If we had authentic presentations of all numbers, as we do of the first few in the number series, there would be no arithmetic, for it would be completely superfluous. The most complicated relations between numbers, which are now discovered only with effort, by means of intricate calculations, would be present to us along with the number presentations and would possess the same intuitive evidence as propositions such as 2 + 3 = 5. For anyone who knows what 2, 3, 5 and the signs + and = mean, this proposition is immediately and evidently clear (PA 191).

Now the phrase 'immediately and evidently clear' might be somewhat questionable here, for in fact we are dealing with an evidence achieved only through eidetic variation over concretely presented numbers. Nonetheless there is plainly a good sense to what Husserl is saying. We can 'see' that the number designated by the sign-complex '2 + 3' is necessarily the same as that designated by the sign '5.' But the situation is quite different if we encounter sign-complexes such as '18 + 49' or '7 x 36.' Here we face something that is truly 'problematic': 'how to decide whether any two of these are equal or not, and then which is the greater and which the smaller?' (PA 261). According to Husserl, it is the arithmetician's business to know the answers to such questions.

In everyday life, of course, we ordinarily write the number signs on paper and proceed to 'calculate' the result in mechanical fashion. We transform the original number names into mere signs, derive another sign from these by following certain established rules, then take the resulting sign as a name once again (PA 257 f.). What the derived name names is the sought-for number. But the scientific arithmetician cannot be satisfied with this 'sedimented' procedure.[35] For him, mere opinion about number relations is not sufficient. He seeks to verify the rules taken for granted in calculation by a return to the things themselves. His focus therefore cannot be on mere number signs. It is for this reason that Husserl distinguishes between the 'art of calculating' and the 'art of *arithmetical* knowledge' (PA 259). The arithmetician must return to numbers themselves, the very numbers that are named by number signs. It is only because numbers themselves bear certain relations to each other that the signs used to designate them can constitute the coherent and reliable system that they do. 'Accordingly, in order to provide a foundation for the *arithmetical calculating methods*, we will have to go back to the *number concepts* and their *connections*' (PA 259).[36] The arithmetician is thus the custodian of the truths at work in everyday calculation; he actually registers what we ordinarily possess only in sedimented, inauthentic form.[37]

Of course, the arithmetician cannot form fully authentic presentations of large numbers any more than others can. Large numbers are never given in the manner of small numbers, just as three-dimensional objects are never given in the manner of two-dimensional surfaces. But the arithmetician can achieve a more authentic presentation of large numbers than that which we have when we calculate with the help of written signs. Instead of 'adding' two 'numbers' by manipulating the corresponding numerals, he can go back to the determinate multitudes which the signs designate. He can actually run through first one multitude, then the other, in one continuous act of counting (PA 264). It is only if this activity is to be called 'calculation' that we can identify 'the art of calculating' with 'the art of arithmetical knowledge.'[38]

These considerations allow us to understand what Husserl means when he says in PA that 'the sense and purpose of the arithmetic of numbers (*Anzahlenarithmetik*)' is determined by the fact that 'we are almost completely limited to symbolic number concepts' (PA 7; cf. 190). Arithmetic is concerned to verify the truths presupposed in ordinary calculation; but if we were able to have fully authentic presentations of all numbers and their relations, such calculation would not be necessary, and hence arithmetic itself would be superfluous. Now this is a thoroughly plausible point of view. Nonetheless it requires some qualification if it is to be made acceptable in the context of Husserl's later interpretation of arithmetic as a branch of formal ontology. In PA itself, Husserl seems not to distinguish sharply between the 'sense' and the 'purpose' of arithmetic.[39] He does not call attention to the difference between the theoretical and the practical sides of arithmetic. He implies that setting up rules and procedures for mechanical calculation – even to the point of deciding what materials should be used for writing the calculation signs – is just as much a part of arithmetic as is the registration of necessary truths about numbers and number relations.[40] But surely there is a fundamental distinction to be made here. Husserl himself calls attention to this distinction in LU. There is a difference, he says in this later work, between pure, theoretical arithmetic and the 'practical art of calculating,' just as there is between geometry and the practical art of surveying (LU I 38 [39 f.]). A 'law' in the sense of a necessary truth about number is not the same as a 'rule' of calculation: the former makes 'no reference to our judging and the manner in which it *should* be performed' (LU I 156 [169]).

Even in LU Husserl admits that it is illuminating in a certain respect to know what logical needs arithmetic answers to; but we would be in error, he suggests, if we took a specification of these needs for a full clarification of the sense of arithmetic.[41] Arithmetic is an eidetic science of number, a branch of formal ontology. Through it are disclosed essential truths about a certain region of objects, the formal region of ideal numbers. This clarification of the sense of arithmetic is of great systematic importance in the context of Husserl's philosophy of arithmetic as a whole. For, as we shall see in the following chapter, the sense of the 'higher' mathematical discipline known as analysis can be clarified only in light of the 'ontological' character of the original arithmetic from which it arises.

NOTES

1 PA 7; cf. above, p. 1 ff.

2 E.W. Beth, *The Foundations of Mathematics* (Amsterdam: North Holland Publishing Co., 1968), p. 353.

3 On Plato's concept of number, see especially Klein, *Greek Mathematical Thought*, pp. 46-60, 69-99. Plato is said to have distinguished between three kinds of numbers: sensible numbers, mathematical numbers and eidetic numbers; cf. Aristotle, *Metaphysics* 987 b 14 ff., 1090 b 35 ff. In each case 'number' means 'a definite number of definite things'; the number kinds are distinguished only in accordance with the kinds of units they contain. Cf. Klein, *Greek Mathematical Thought*, pp. 46 ff. In sensible numbers, the units are themselves sensible things, while in mathematical numbers the units are 'pure' monads, outside of time and change. The units of the eidetic numbers, at least on Klein's interpretation, 'are nothing but a conjunction of *eide* which belong together'; for example, 'the eidetic 'two' ... represents the *genos* of '*being*' *as such*, which comprehends the two *eide* 'rest' and 'change'' (ibid., pp. 89 f., 93). Now it is possible to find something in Husserl's philosophy analogous to each of these number kinds. To Plato's sensible numbers there correspond Husserl's 'concrete' or individual numbers, through of course Husserl stresses that not even these numbers are actually given through sense perception. To Plato's mathematical numbers there correspond Husserl's 'ideal' numbers, which are also understood as outside of time and motion and as the object of the arithmetician's study. There seems to be an important difference between Plato and Husserl, however, in that Husserl views ideal numbers as themselves the unitary 'species' or 'essences' of the corresponding concrete numbers, while Plato apparently regards the mathematical numbers as being 'many and alike'; cf. Aristotle, *Metaphysics* 987 b 15 ff. In other words, for Plato there would seem to be 'many' mathematical Two's, but for Husserl there is only one ideal Two. Perhaps it is our ability to say truthfully that 'two and two are four' which leads Plato to suppose that there are many mathematical Two's. If so, the Husserlian (and Fregean) distinction between the *meaning* (or sense) of an expression such as 'two and two' and the *object meant* (or referent) could be said to make this supposition unnecessary: 'two and two' differs in sense from 'four,' but not in reference. But by far the most difficult problem in correlating Husserl's concept of number with that of Plato concerns the latter's eidetic numbers (*arithmoi eidetikoi*). There seems at first to be nothing of this sort in Husserl. It is true that just as Plato limits the eidetic numbers at ten (cf. Aristotle, *Metaphysics* 1073 a 10, 1084 a 12 ff.; *Physics* 206 b 32 f.), so too Husserl limits the 'authentic number concepts' at 'twelve (or some smaller number close to it)' (PA 192; cf. 222); cf. G. Martin, *Klassische Ontologie der Zahl, Kantstudien*, Ergänzungsheft 70 (1956), pp. 17-29 and 142 f. But these limitations appear to arise from entirely unrelated considerations. R. Sokolowski however suggests another, more profound sense in which Husserl's philosophy could be said to contain something analogous to Plato's

concept of eidetic number: just as 'being' for Plato is the eidetic two, rest and change taken together, so in Husserl's philosophy we might say that 'being' is found in the 'dyad' of presence and absence and the 'one' they make together; see 'Ontological Possibilities in Phenomenology: The Dyad and the One,' *Review of Metaphysics* 29 (1976), pp. 691 ff. In other words, 'being' is understood to have an '*arithmos* structure' in Husserl and Plato alike. For Husserl as well as Plato, being is an eidetic number.

4 This is the definition of 'Platonism' given by P. Bernays in 'On Platonism in Mathematics,' in *The Philosophy of Mathematics. A Book of Readings*, ed. by P. Benacerraf and H. Putnam (Englewood Cliffs, N.J.: Prentice-Hall, 1964), p. 275.

5 Husserl does speak from time to time in PA of what he terms 'numbers in themselves' (*Zahlen an sich*). In one of the most prominent passages where this occurs, he also uses the phrase 'idealizing abstraction' (*idealisierende Abstraktion*) (PA 260). The superficial resemblance between these phrases and certain others which occur in LU might lead one to suppose that the Husserl of PA was very close to recognizing 'ideal numbers,' as he does explicitly in LU. In fact, however, this is not the case. The 'idealizing abstraction' referred to in PA is by no means the same as the 'ideational abstraction' described in LU. As PA 260 makes clear, this phrase simply refers to the disregarding of the factual limits on our ability to count; it does not mean finding a unity in a class of concrete numbers. The 'numbers in themselves' of PA are simply concrete numbers or number concepts in contrast to the signs (or 'systematic numbers') which represent them. They are not ideal objects.

6 See above, pp. 39 ff.

7 Husserl presents this theory as one taught by certain unnamed 'logicians,' but he clearly quotes it with approval. The complete passage reads as follows: 'Wie gelangt man denn ... von dem Konkreten zum Abstrakten? Die Logiker lehren, jeder Allgemeinbegriff zu seiner Entstehung voraussetze irgendeine konkrete Anschauung. Das Abstractum entsteht nun derart, daβ aus der Gesamtanschauung des Konkreten durch die isolierende Aufmerksamkeit ein gewisses Moment, eine Seite, Eigenschaft oder Beschaffenheit hervorgehoben word. Und eben dieses Moment ist es, das alle Konkreta des Begriffs gemeinsam haben und um dessentwillen wir sie eben alle Gegenstände des Begriffs erklären' (Ms. K I 28/5b). For confirmation of the view that Husserl did in fact adhere to this theory of abstraction, see PA 79, where it is said that 'to disregard or abstract from something means simply: not to notice it especially.'

8 PA 78; emphasis added.

9 Husserl's early theory of abstraction had been strongly criticized by Frege in his review of PA; cf. Frege, 'Rezension,' p. 181 f.

10 Husserl himself classifies 'concept' in LU among the 'categories of meaning'; 'number' on the other hand is said to be an objective category (LU I 243 f. [237]). On Husserl's distinction between concept and essence, see Sokolowski, *Husserlian Meditations*, pp. 68 ff.

11 Cf. LU II/1 72 (307): 'Where there is talk of knowledge 'springing from the analysis of the mere meaning of words,' more is meant than these words suggest ... Such analysis is not ... concerned with empty thought-intentions, but with the objects and forms by which they are fulfilled. What it therefore offers us are not mere statements concerning elements or relations of meanings, but evident necessities concerning the *objects* thought of in these meanings, and thought of as thus and thus determined.'

12 Husserl says at one point in LU that when we make the ideal Four authentically present, 'some individual group of four units certainly floats pictorially before us, and to that extent underlies our presentation and our judgment' (LU II/1 141 [366]). This should not be interpreted to mean however that categorial activity is inessential; we can have a sensuous group 'pictorially before us,' but not a multitude of four as such.

13 There is no mention in LU of the view that numbers arise by 'eliminating the indeterminateness' in the abstract concept of multitude, as Husserl had suggested in BdZ and in the corresponding passages of PA. His position in LU is much closer to the second of the two accounts of the origin of number concepts presented in PA (that according to which number concepts are abstracted directly from the corresponding concrete numbers).

14 See above, pp. 72 ff.
15 The terminology here is, to be sure, somewhat confusing. Elsewhere in LU, ideational abstraction *in general* is characterized as a founded, 'categorial' act (cf. LU II/2 161 ff. [799 ff]). This would seem to mean that sensuous abstraction is categorial as well as categorial abstraction itself. The difficulty is cleared up however if we bear in mind that though the act of abstraction itself is always categorial, the acts on which it is founded are not.
16 Cf. Tugendhat, *Wahrheitsbegriff*, p. 141 f.; Sokolowski, *Husserlian Meditations*, pp. 61 ff.
17 Despite his otherwise illuminating discussion of Husserl's doctrine of abstraction, Tugendhat has seriously misrepresented Husserl's view on one point. Referring to the doctrine of categorial abstraction, Tugendhat comments: 'Es gibt demnach auch eine 'Abstraktion,' die nicht in der Hervorhebung eines inhaltlichen Momentes im vorgegebenen Gegenstand besteht, sondern in der pointierenden Reflexion auf die konstituierende kategoriale Synthesis, so dass die konstituierte Gegenständlichkeit spezifisch *als* Produkt dieser Synthesis bei beliebiger sinnlicher Grundlage erscheint' (*Der Wahrheitsbegriff bei Husserl und Heidegger*, p. 147). In other words, categorial abstraction as Tugendhat understands it consists not in thinking about what is common to a manifold of individual categorial *objects*, but rather in reflecting on the synthesizing *act* which constitutes such objects. But this is plainly a misrepresentation of Husserl's doctrine, as we see clearly from the latter's discussion of the concept of 'identity': 'If we look at an intuitive act of identification, i.e. an intuition of identity, and if we abstract from it the moment of *identification* (*Identifizieren*), we have performed a sensuous abstraction. But if, *while living through an identification*, we turn our regard to *objective identity*, and make *this* the basis of our abstraction, we have performed a categorial abstraction ... The objective moment of identity is no act, and no form of an act; it is an objective categorial form' (LU II/2 178 f. [814]). Likewise, in discussing the abstraction of the ideal Five, Husserl says that we 'not only isolate the non-independent moment of collective form in what is before us, but we apprehend the Idea in it' (LU I 171 [180]).
18 See also *Ideen* 26 ff. (59 ff.).
19 FTL 77 (82); cf. *Ideen* 28 (61).
20 Husserl distinguishes 'something' from 'one' on the grounds that 'one' has a 'correlation to multitude as a connotation' (PA 84). 'Something,' Husserl implies, has no such correlation; it is rather prior to the concept of multitude.
21 Cf. *Ideen* 26 f., 32 ff. (60, 64 ff.).
22 In FTL Husserl stresses that categorial forms, the 'modes of anything whatever,' accrue in acts of judging that are still straightforwardly object-directed; cf. FTL 119 (113 f.). This seems to imply that the doctrine presented in PA regarding the origin of the concept of 'something' was incorrect.
23 PA 16. The views criticized by Husserl include those of Aristotle (number is a 'common sensible'), certain Scholastics (number arises through a division of the continuum, and is therefore inapplicable to incorporeal beings), Locke (number is a 'primary quality') and Mill (number is 'physical property,' a 'visible and tangible' phenomenon); cf. PA 16 f., 85.
24 In some parts of PA, Husserl himself suggests that, historically and psychologically speaking, the purely formal concept of number is not the most 'original' one (PA 44, 146 f.). He seems however to attach no philosophical significance to this point.
25 EU 394 ff. (327 ff.). See also Sokolowski, *Husserlian Meditations*, p. 8 ff.
26 A striking example of a culture that evidently had not arrived at a formal concept of number is given by T. Dantzig in *Number: The Language of Science* (New York: MacMillan, 1954), p. 6: he reports that in the language of a certain British Columbian tribe, 'we find seven distinct sets of number words: one for flat objects and animals; one for round objects and time; one for counting men; one for long objects and trees; one for canoes; one for measures; one for counting when no definite object is referred to.' Dantzig adds that the last set of number words is 'probably a later development.' See also B. Russell, *Introduction to Mathematical Philosophy* (London: Allen and Unwin, 1967), p. 3: 'It must have required many ages to

discover that a brace of pheasants and a couple of days were both instances of the number 2: the degree of abstraction involved is far from easy.' Husserl himself claims in FTL that even the ancient Greeks lacked a purely formal concept of number: 'not even their concept of cardinal number is emptied of all non-formal material; in the units thought of as counted, it is not yet related to the realm of the empty anything-whatever' (FTL 83 [80]).

27 *Ideen* 27 (60). The term 'ontology' is not used in LU, but as Husserl points out in a footnote in *Ideen*, the idea of an '*a priori* theory of objects' is nonetheless implicit in the work (*Ideen* 28 n. [61 n.]; cf. LU II/1 225 f. [435]).

28 *Ideen* 23 (57); FTL 81 f. (77 f.).

29 LU I 244 (237). In *Ideen* Husserl claims that formal ontology 'in the full inclusive sense' would embrace apophantic logic as well, since 'pure truths concerning meaning can be transformed into pure truths concerning the object' (*Ideen* 28 [61]). But he goes on to say that 'nonetheless the meaning-categories must be separated off as a group having its own distinctive character, and the remaining ones set over against them as the formal objective categories in the pregnant sense of the term' (ibid.).

30 Husserl claims in FTL of course that 'formal ontology and formal apophantics, despite their different themes, must be very intimately related and are perhaps inseparable' (FTL 83 [78 f.]). Both belong, he claims, to one comprehensive '*mathesis universalis*' (FTL 53, 78 ff. [48, 73 ff]). On this question see E. Winance, 'Logique, mathématique et ontologie comme 'mathesis universalis' chez Edmond Husserl,' *Revue Thomiste* 66 (1966), pp. 410-434; and Sokolowski, *Husserlian Meditations*, pp. 271-289.

31 See above, p. 90. Husserl seems to be aware even in PA of the difficulties which attend the view that numbers themselves are 'concepts.' Despite the fact that there *are* number concepts, he says at one point that these concepts are not what the arithmetician studies: in arithmetic '5 does not mean the concept (the abstractum) five; 5 is rather a general name (or a calculating sign) for any arbitrary set falling under the concept five as such' (PA 181 f.). In this passage Husserl recognizes that arithmetic studies objects as distinct from concepts, but is as yet unable to see that the objects in question are in fact *ideal* objects.

32 Given the aforementioned similarities between the Husserlian and Platonic concepts of number (see above, p. 104 n. 3), it is interesting to note that Husserl's concept of arithmetic *differs* significantly from that of Plato. What Husserl calls 'arithmetic' seems to correspond most closely to what Plato calls 'logistic,' and not to what Plato calls 'arithmetic'; cf. *Gorgias* 451 A-C and Klein, *Greek Mathematical Thought*, pp. 17-25. According to Klein, Platonic arithmetic was a study of the kinds or *eide* of number, such as the odd, the even and so on (ibid., pp. 54-60). Klein holds that this study was made necessary at least in part by the problem of the 'unity' of number: 'precisely because the *arithmos* as such is not one but many, its delimitation in particular cases can be understood only by finding the *eidos* which delimits its multiplicity, in other words, by means of *arithmetike* as a theoretical discipline' (ibid., p. 56). Husserl, as we have seen, also recognizes the unity of number as a problem, but he deals with it through phenomenological analyses and therefore does not need an 'arithmetic' of the Platonic variety; see above, pp. 69 ff.

33 See above, pp. 55 ff.

34 See above, p. 76 f.

35 On the 'sedimented' character of calculation, see above, pp. 81 ff.

36 When Husserl writes that 'the method of sensuous signs is thus *the* logical method of arithmetic' (PA 257), he is referring to the calculating technique which the arithmetician's insight founds, not to the arithmetician's own activity. The arithmetician himself must go beyond the signs to numbers themselves. As Husserl would have put it later in his career, the arithmetician must return to the ideal numbers given only in eidetic variation over concrete numbers.

37 Cf. Klein's discussion of 'theoretical logistic' in Plato; *Greek Mathematical Thought*, p. 22 f.

38 PA 257; cf. 264. The 'essentially conceptual operation' described here is distinguished by Husserl from three other conceivable senses of the term 'calculation.' First, there is the combination of two authentically presented numbers to form a third number also authentically presented (PA 256). Since no 'method' is involved here, it seems rather implausible to call this activity calculation. Second, there is the 'essentially sensuous operation which on the basis of the number sign system derives signs from signs according to fixed rules, then takes the result as the designation of a certain concept, the concept that was sought' (PA 257). This, Husserl says, is the 'most common' concept of calculation (PA 257 f.). Finally, there is the derivation of signs from signs *without* any subsequent interpretation of the result as the name of a sought-for concept (PA 258). This sense of calculation arises only from 'higher logical interests than those of the arithmetica numerosa.' On this latter sense of calculation, see below, p. 115.

39 Husserl says at one point in the work that we can clarify the 'true sense' of arithmetic by explaining the 'logical needs which it arises from and satisfies' (PA 272).

40 Cf. PA 243 f., 262, 272 ff.

41 In Chapter IX of the Prolegomena, Husserl suggests that a detailed study of the way symbolic methods enable us to overcome the limitations on our intellectual powers is an important part of logic in the broadest sense of the term; see especially LU I 197 ff. (200 ff.), including the reference to PA at LU I 202 n. (204 n.). Husserl is concerned in this chapter with the principle of 'economy of thought' and its relation to the foundations of logic. He contends that this principle is 'important and fruitful,' 'an important foundation' for logic in the practical sense of a 'technology of knowledge' (*Kunstlehre der Erkenntnis*) (LU I 202 [204]). He denies only that it can serve as the foundation for *pure* logic: 'pure logic is in itself prior to all such thought-economics, and it remains absurd to base the former on the latter' (LU 209 [209]).

CHAPTER VI

THE SENSE OF ANALYSIS

1. THE ALGEBRAIZATION OF ARITHMETIC

In the previous chapter we saw that Husserl interprets arithmetic, the original science of number, as a branch of formal ontology. It is an 'ontology' in the sense that it aims to establish essential truths about a region of objects, and it is 'formal' in the sense that the objects in question are purely formal objects. Ultimately, as we have seen, arithmetic rests on the evidence obtained through eidetic variation over the concrete numbers constituted in acts of collecting and counting. Now Husserl's clarification of the sense of arithmetic undoubtedly possesses a certain interest in its own right. Systematically, however – that is, in the context of Husserl's 'philosophy of arithmetic' as a whole – this issue is a subordinate one. We saw in Chapter I that Husserl's ultimate concern was not with elementary arithmetic as such, but rather with the higher mathematical discipline known as analysis.[1] He hoped to contribute to the 'true philosophy of the calculus' by clarifying and justifying modern analysis, the mathematical discipline in which he himself had received his doctorate. We must therefore proceed to examine his work in this area. Just how did Husserl attempt to clarify and justify analysis? We shall take up the issue of justification in the final section of this chapter. In the first three sections, we shall be concerned solely with Husserl's attempt to clarify the sense of analysis.

As we shall see, the sense of analysis, as Husserl interprets it, is quite different from the sense of the original arithmetic. Despite the fact that analysis is essentially concerned with what is usually called the 'real number system,' this discipline is *not* founded on the evidence generated through eidetic variation over concretely presented numbers. The intentional structures at work in analysis, in other words, are essentially different from those at work in arithmetic. Husserl expresses this difference by saying that analysis, in contrast to the original arithmetic, is a branch of 'manifold theory' (*Mannigfaltigkeitslehre*). Our task in the first sections of this chapter is to explain the meaning of this claim.

See notes at the end of this chapter.

The first issue we must take up concerns what might be called the 'algebraization of arithmetic.' In his late work *Die Krisis der europäischen Wissenschaften*, Husserl calls attention to 'the enormous effect – in some respects a blessing, in others portentous – of the algebraic terms and ways of thinking that have been widespread in the modern period since Vieta ...' (*Krisis* 40 [44]). But this concern with the nature and significance of algebraic thinking was by no means something that Husserl developed only late in his career. On the contrary, it had been a central component of his philosophy of arithmetic from the very beginning of his work in this area.[2] In the manuscript of his 1887 lectures on the philosophy of mathematics, for example, we already find an allusion to the fundamental importance of an 'algebraic sign language.'[3] Husserl describes this as the first great discovery of modern mathematics, the discovery which made all subsequent developments possible.[4] It was this discovery, he suggests, which generated the distinctively modern form of mathematics: it led to 'theories and methods ... possessing a generality and significance which the Greek mathematical spirit never suspected or strove for.'[5] In FTL Husserl also calls attention to this development; he says in fact that 'the genuine discovery of the formal was first made, at the beginning of the modern age, by way of Vieta's establishment of algebra ...' (FTL 84 [80]). But this formulation is somewhat misleading. The concept of 'the formal' which algebra made possible must not be confused with the formality inherent in the concept of number itself. The latter arises, as we saw in the previous chapter, through eidetic variation over concrete numbers. It can on principle be attained prior to the introduction of algebraic techniques.[6] The concept of the formal to which algebra leads, on the other hand, is a formality of a wholly different kind. It is a level of formality on which the original arithmetical thinking 'becomes free, systematic, *a priori* thinking, completely liberated from all intuited actuality, about numbers, numerical relations, numerical laws' (*Krisis* 40 [44]). Algebraic formality is not inherent in the concept of number; rather, it consists in a 'liberation' of arithmetical thought from the concept of number.

How does the introduction of algebraic techniques lead to such a liberation? The first step is the development of what Husserl terms 'universal arithmetic.' In universal arithmetic, we employ literal symbols to express truths about *ideal numbers in general*. As Husserl notes in *Ideen*, there are levels of generality within the realm of number just as there are in other regions. 'Number in general' (*Anzahl überhaupt*) is a 'highest genus,' while 'two, three, and so forth are its lowest differences or eidetic singularities.'[7] We saw in the previous chapter that the original arithmetic is concerned to register truths about the relations between these 'eidetic singularities'; that is, it is concerned with truths about the relations between the various individual ideal numbers. On this level we might, for example, register the truth that $7 + 5 = 5 + 7$. In universal arithmetic, however, we would be concerned with the more general truth that $a + b = b + a$, where *a* and *b* stand for any numbers whatever.[8]

Why are such general arithmetical truths of interest? Husserl discusses this question in the final chapter of PA. One reason for universal arithmetic is the desire to simplify complicated problems of calculation, and thereby save ourselves 'time and energy' (PA 279 f.). It is useful to know, for example, that the result of multiplying a given number by a second number, then dividing by the second number, is necessarily equal to the very number that was given originally; in other words, that $\frac{a \times b}{b} = a$. A second reason for universal arithmetic is the need of solve equations involving 'unknown' numbers (PA 281 f.). It sometimes happens that a problem of calculation is indicated only indirectly, as when we are told that the result of multiplying a given number by an 'unknown' number, then adding a second number, is equal to a third number. It is obviously quite useful to express such problems 'algebraically': $ax + b = c$. Only when we adopt this way of formulating such a problem do we find it plausible to state the value of the unknown number in the 'general' form: $x = \frac{c - b}{a}$. Having derived this new equation, we are in position to determine whether the original equation did in fact have a numerical solution, and if so what the most efficient means of calculating that solution would be.[9]

The use of letters to stand for numbers in general also makes it possible to present the whole complex of arithmetical truths in more systematic form. It becomes possible, as Husserl says, to give arithmetic the unity and systematic coherence of a deductive theory, analogous to that of Euclidean geometry.[10] Some truths can be taken as axioms, others as theorems to be derived from these by deductive inference. Husserl himself elaborates a set of such axioms in one of his early manuscripts.[11] But it is important to stress that at this stage of development, the axioms are not mere arbitrary assumptions. The axioms on which universal arithmetic is based are themselves necessary truths or 'laws' pertaining to numbers in general. The proposition '$a + b = b + a$' asserts a relation that necessarily holds between any two numbers. In the manuscript just alluded to, Husserl says that the axioms he proposes are 'immediately evident' (*unmittelbar einleuchtend*).[12] Perhaps this is not an altogether satisfactory way of expressing the matter, for the evidence in question is in fact obtained only through a process of eidetic variation. Nonetheless what he says does have a good sense insofar as it brings out the non-arbitrary character of these axioms. As Husserl remarks in LU, 'the propositions of universal arithmetic (*arithmetica universalis*) ... are laws rooted in the ideal essence of the genus number' (LU I 172 [181]). There is of course some room for choice in regard to which laws are laid down as axioms, which derived from these as theorems. But we are not at liberty to set forth as an axiom a proposition which is not universally and necessarily true of numbers. We might want to set down as an axiom the proposition that '$a + b = c$'; that is, that the sum of any two numbers is itself a number. But we could not possible assume '$a - b = c$,' for the expression '$a - b$' designates a number only in cases there *a* is greater than *b*.[13] We cannot substract a greater number from a smaller. All of the 'inverse operations,' as Husserl calls them, are subject to similar limitations.[14]

At the stage of development we have been considering, the new sense of 'the formal' referred to earlier in this section has not yet emerged. Universal arithmetic represents only a 'generalization' of the original arithmetic, not a 'formalization.' The use of algebraic symbols and techniques has not yet led to a shift in the intentional structures at work in arithmetic. We are still intending ideal numbers, determinate multitudes of units. The only difference is that we now register truths that relate simultaneously to *all* such numbers.[15] The number symbols that we use (a, b, c, ...) are simply general number names rather than particular ones. Nonetheless, Husserl claims, there is in algebraic calculation 'an inherent tendency toward formalization' (PA 432). The use of algebraic number signs

> leads imperceptibly to formal generalizations of our original thought-trains, and even of our sciences. In this manner, almost without specially directed mental labour, deductive disciplines arise having an infinitely enlarged horizon. Out of arithmetic, the original theory of numbers and numerical magnitudes, a generalized, formal arithmetic arises in more or less spontaneous fashion: in this numbers and magnitudes no longer count as basic concepts but merely as chance objects of application (LU I 200 [203]).

It is only at this point, when arithmetic has been 'liberated' from the concept of number, that a genuinely new kind of formality emerges.

The step which leads to algebraic as distinct from numerical formality is explained in somewhat greater detail in a manuscript dating from 1907.

> This form first emerges in its purity in the mathematical method, i.e. by means of algebraic relations and the sensuously external forms of connection ... It can quite easily happen – and does happen often enough – that two mathematical theories established in quite different material regions agree completely in regard to form. This happens when the two theories proceed from fundamental principles and concepts which are of altogether similar formal constitution, even though in regard to content they have different meanings. No matter whether the letters mean, for example, sets or numbers, and '+' the combination of sets into further sets or that of numbers into further numbers, ... the proposition 'a + b = b + a' still holds. We can also understand the letters as line segments and the plus-signs as the adjoining of one line segment to another ... Once again 'a + b = b + a' is valid. The proposition has a different sense, ... but in form it agrees completely with the previous one and thus receives the same algebraic expression.[16]

In such a situation, Husserl goes on to note, 'it is clear that one can liberate the form of the mathematical system from its region, from its matter, so to speak ...'[17] We can retain the symbols and the various propositions formulated by means of them even while ceasing to intend any one of the regions to which the propositions in question originally referred. We can simply take 'a + b = b + a' as a given, and then proceed to see what follows from it. As Husserl says, one does not need actually to carry out the deductions twice.'[18]

In this shift of focus, which only becomes possible with the introduction of algebraic methods, a new sense of formality is generated which is obviously quite different from that at work in the original theories themselves. Arithmetic and set theory were formal in the sense that their objects (the numbers and sets theorized about) were formal. But the new discipline is formal in the sense that it studies the 'theory form' that is common to several different actual or possible theories. We are now studying not what is common to a class of *objects*, but rather what is common to a class of *theories*. It is only in contrast to this new sense of formality that even numbers and sets can be said to make up 'material regions.' We have moved from numerical formality to algebraic formality, from 'formal ontology' to the study of a 'theory form.'

Given this new mathematical orientation, in which our primary concern is no longer that of developing an actual theory of a given object region, the difference between one theory form and another can come to see more or less arbitrary. The difference between a formal system in which we restrict the use of the symbol '–' (allowing 'a – b' only where it is established that 'a > b'), and another system in which we do not, can seem only to reflect a choice someone has made. The axioms assumed come to seem mere 'conventions' which can plausibly be subjected to variation. Different axioms, we come to see, would generate entirely different theory forms, some of which might be entirely free of internal contradiction. Thus there arises the idea of a new science, a science which 'deals *a priori* with the *essential sorts (forms) of theories and the relevant laws of relation*' (LU I 247 [240]). Husserl calls the new science 'the theory of possible theory forms' (*die Theorie der möglichen Theorienformen*) (LU I 247 [239]).

In Husserl's view, an adequate understanding of the sense of analysis presupposes an understanding of the intentional structures at work in the theory of theory forms. We shall examine these structures in the section that follows.

2. THEORY FORMS AND MANIFOLDS

When we shift from the study of an object region to the study of theory forms, we make a move into what Husserl calls the 'apophantic' domain, the domain of meanings as distinct from objects meant. In the original ontological theories (arithmetic and set-theory, for example), our primary aim was of course to formulate truths about the regions in question. However, we were also concerned to present these truths systematically. We were systematic in that we sought both to exclude contradictions and to derive some judgments from others by deductive inference. We were concerned, in other words, with consistency and consequence as well as with truth. Now in the shift to the study of theory forms, this concern undergoes an important modification. We begin to focus on theories as such, on theories simply as 'judgment-systems' (FTL 94 [90]), disregarding their claim to present truths about a region of objects. The concern for truth is 'bracketed,' so to speak; consistency and consequence alone are our themes. In this sense, the attitude which we take in the theory of theory forms is identical with that of traditional formal logic, or with what Husserl calls 'apophantic analytics' (FTL 76 f. [72 f.]). Husserl says in fact that the theory of theory forms is to be considered the 'highest level' of formal logic.[19] In both disciplines – in the theory of theory forms as well as in traditional logic – we are concerned with the *relations between* judgments, not with the *correctness of* judgments. The intentional structures at work in the two disciplines are therefore comparable. But Husserl calls attention to an important distinction that must be made regarding the intentionality at work in the theory of theory forms. This distinction – which does not seem to be needed on the 'lower' level of apophantic logic, the level corresponding to the traditional theory of inference[20] – concerns the attitude that is taken to the formal systems that are studied. Two such attitudes are conceivable. One of these generates what Husserl calls the 'mathematics of game-rules,' the other what he calls 'manifold theory.' We must examine this issue carefully, for it is crucial to Husserl's clarification of the sense of analysis.

The mathematics of game-rules (*Spielregeln*) arises if we come to regard the study of theory forms as an end in itself, without any broader epistemic significance whatsoever. Where this attitude takes hold, we become thoroughly absorbed in the activity of manipulating symbols; as Husserl says, we become 'lost in an excessive symbolism' (FTL 104 [99]). Instead of regarding the study of theory forms as a serious scientific enterprise, we view it as 'only a *discipline comprising deductive games with symbols*' (FTL 104 [99]).

> One operates with letters and with signs for connections and relations (+, x, =, etc.), according to the *rules of the game* for arranging them together in a way not essentially different, in fact, from a game of cards or chess (*Krises* 46 [46]).

Now it would be wrong to say that this game-mathematics involves operation with 'meaningless signs.' As Husserl points out in LU, even in symbolic thought we 'live entirely in the consciousness of meaning' (LU II/1 68 [304]). It is true of course that the signs we use are no longer taken in their original 'arithmetical meaningfulness' (LU II/1 60 [305]). Our letters do not signify numbers, nor do our operation signs signify connections between numbers. Nonetheless the signs in question do have a meaning of sorts for us; they are something more than mere physical objects, mere marks on paper. As Husserl puts it, they have a 'games-meaning' (*Spielbedeutung*); that is to say, they have not their original meaning but 'a meaning oriented toward the game of calculation and its well-known rules' (LU II/1 69 [305]). The 'meanings' of these signs are understood to be derived from the axioms of the formal system in question, just as the 'meanings' of the pieces used in playing chess are derived from the rules of chess. The bishop, the rook and the other pieces are not just 'bits of ivory and wood,' as Husserl points out. 'They become chessmen, counters in the chess-game, through the game's rules which give them their fixed *games-meaning*' (LU II/1 69 [305]). To take a rook as a rook is to take it as something for which certain 'moves' are permitted in the game of chess. When we are conscious of the permitted moves, we may be said to be conscious of the piece's meaning. And in the mathematics of game-rules, the algebraic symbols used are taken as meaningful in just the same way.[21] Though not utterly bereft of meaning, their meanings are understood simply as the totality of moves permitted for them within the system. They are taken to have a sense, so to speak, but no reference; to have meaning, but a meaning that lacks all relation to an object meant.

Husserl alludes to this game-mathematics already in PA. He mentions that sign-equivalences can be established as mere conventions, 'in the manner of game-rules' (PA 239). And although he fails to see that even the counters in a game are somehow meaningful (cf. PA 237), he does nonetheless draw an interesting comparison between the symbolic thinking at work in game-mathematics and that at work in ordinary arithmetical calculation.

> In light of the fact that the mechanism of the symbolic method can be fully separated from the conceptual substrates to which it applies, still another concept of calculation comes to the fore ...
>
> One can take it namely as every rule-governed manner of deriving signs within an algorithmic sign-system, according to the 'laws' – or rather: conventions – of connection, separation and replacement which are peculiar to this sign-system.[22]

When we manipulate the signs of an algorithmic system in this way, we are absorbed in the signs that are actually present, just as we are in everyday calculation with number signs. In both cases, the original meaning of the signs we use is a matter of indifference; the only thing that concerns us is the set of manipulations that we are allowed. The difference between the two is primarily a matter of degree: in ordinary calculation, the reductive focus is merely temporary, prevailing only until we arrive at the sought-for result. In the mathematics of game-rules, on the other hand, the shift of focus can become so complete that we actually become 'lost' in symbolism. The original significance of the symbols we manipulate can come to be not merely disregarded for the moment, but actually forgotten.

The attitude which generates game-mathematics is one possibility in the study of theory forms. However, as we noted, another attitude is also possible, the attitude which leads to 'manifold theory.' It is important to stress that manifold theory is not identical with game-mathematics. The intentional structures at work are not the same in the two cases. The difference between them is a subtle one, to be sure; nonetheless we cannot understand Husserl's clarification of the sense of analysis unless we are clear about this difference.

We must note in the first place that a 'manifold' (*Mannigfaltigkeit*)[23] in the sense relevant at present is not simply a 'set (*Menge*), nor is 'manifold theory' an instance of 'set theory.'[24] It is true of course that some authors use the terms 'manifold' and 'set' as equivalents. Husserl however does not.[25] A manifold is not simply a collection of objects, but rather a *region* of objects having certain specified relations to each other. Husserl defines a 'manifold' as 'a region' (*Gebiet*) which is uniquely and solely determined by falling under a theory of [a given]) form, whose objects are such as to permit of *certain* associations which fall under *certain* basic laws of this or that *determinate* form (here the only determining feature).'[26] Now this definition certainly requires elucidation. Nonetheless we can already see that there is an essential difference between a set and a manifold. Manifolds are not constituted in straightforward acts of collecting, as sets are. Rather, they presuppose a move into the apophantic domain. They presuppose that a theory form has first been isolated from a class of actual or possible theories. A manifold arises only as the objective correlate of such a theory form.

For a more complete understanding of what Husserl means by a manifold, we must contrast manifold theory with game-mathematics. In the latter, as we saw, we are thoroughly absorbed in the manipulation of the algebraic symbols. But in manifold theory we take those symbols in a different way:

> In the definition of a manifold we have to define not merely in a signitive and calculative manner – for example: 'It shall be allowed to manipulate the given signs in such a manner that the sign b + a can always be substituted for the sign a + b.' Rather we must say: 'There must be among the *objects* belonging to the manifold ... a certain *combination-form* with the *law*-form a + b = b + a ...' (FTL 105 [100]).

In game-mathematics, the algebraic symbols are meaningful only in the way chessmen are meaningful. They have a meaning, but there is no 'object meant.' In manifold theory however 'one regards the games-symbols as signs for actual thought-objects (*Denkobjekte*) ...' (FTL 104 [99]). In other words, we move to manifold theory when we take the algebraic symbols in our theory form as meaningful not just in the manner of chessmen, but in the manner of genuine *names*. In manifold theory the symbols do not merely play rule-governed roles in an algorithm; they are also taken to designate what Husserl calls 'thought-objects.' Likewise, the axioms are not mere game-rules, mere conventions, but rather 'laws' governing an entire region of such objects. In manifold theory, theory forms are taken to have reference beyond themselves; what Husserl calls 'manifolds' are what they are taken to refer to.

It is clear that Husserl regards manifold theory as somehow preferable to game-mathematics.[25] Nonetheless he is not simply recommending this approach; he is also describing and interpreting the thinking that is already at work in a certain form of mathematics. For example, he finds an unmistakeable shift to manifold theory in the non-Euclidean geometries developed in the nineteenth century. Riemann's great achievement consisted in his 'going back ... to the form of a deductive discipline' and then proceeding to vary it freely (FTL 97 [93]). But Riemann and his successors did not merely speak about the theory form which they made thematic; rather, 'they talk about *n*-dimensional spaces, about Riemannian and Lobachevskian spaces ...' (FTL 97 f. [94]). In other words, the mathematicians themselves talk about and intend object regions that are correlated with the theory forms they develop. Husserl's own role is chiefly to clarify this talk. Instead of calling the object regions of non-Euclidean geometry 'spaces,' we should, Husserl suggests, call them 'manifolds' of various types, and do so consistently. In this way the distinctive character of non-Euclidean geometry can be brought out clearly. *Euclidean* geometry – that is, Euclidean geometry 'as Euclid himself understood it' – represented 'the theory of intuited world space' (FTL 96 [92 f.]). Geometry was understood to be an actual theory of an actual objective region. The region in question was taken to have its own distinctive mode of presence, its own mode of 'being' quite independently of the theory directed toward it. With Riemann and his successors, however, a subtle but profound intentional shift came to permeate geometrical thought. The geomatry that had been handed down since Euclid's time underwent a 'formalizing reduction' through which 'all the specifically spatial concepts [were] converted into indeterminates, modes of 'anything whatever'' (FTL 96 [92 f.]). In other words, when Riemann and his successors used the terms 'line,' 'plane,' etc., in their 'geometries,' while making the parallel postulate subject to modification, they were naming not the familiar phenomena of the life-world, nor any 'idealizations' of such phenomena, but rather objects whose specific character was determined in each case only by the axioms of the theory form in question. The object regions they referred to were not regions that could be made authentically, intuitively present. It was not as though they first had insights into the nature of the regions they referred to, then constructed theories which would express truths about these regions in systematic form. Rather, the relation between theory and region was exactly the reverse of this. They thematized the theory form itself first of all, the object region only as something that was somehow correlated with the theory form. As a result,

> what we call axioms in 'non-Euclidean geometry' are in truth not complete propositions, just as the fundamental concepts are in truth not concepts. And thus non-Euclidean geometry is actually not a mathematical science in the ordinary sense, not a theoretical system of propositions grounded in purely deductive fashion on fundamental truths.[28]

Non-Euclidean geometry is an example of a new kind of mathematics – a kind which Husserl proposes to designate as 'manifold theory.'

But it is not only in geometry that a shift to manifold theory has taken place. Something analogous is to be observed in the sphere of arithmetical thinking as well. Here too we find that the reduction of a theory to a freely alterable theory form is coupled with an interpretation of the form as referring to a correlated object region. In this case of course the correlated object region is considered to be a region of 'numbers' rather than of spatial objects. But here again, it seems, we might more appropriately speak of a 'manifold' of a certain kind, in this way we can bring out the distinctive character of the objects thematized in this new approach.[29]

We shall have more to say about the specific manifold correlated with the theory form of arithmetic in the section that follows. At present, we need to examine the general notion of a manifold more closely. Some unavoidable questions arise in connection with this notion. Just what are the 'thought-objects' which make up a manifold? And what leads us to posit them?

The thought-objects in question are objects of an unusual kind. They are not 'ideal objects' constituted through eidetic variation over a class of concrete objects. 'They are not determined as individual or specific singularities, nor indirectly by way of their material species and genera, but solely by the *form* of the connections attributed to them' (LU I 249 [241]). Their phenomenological character is therefore quite different from that of the ideal numbers discussed in the previous chapter. Far from being intuitable essences, these thought-objects have no presence at all apart from the theory form in which they are named. It is for this reason that

Husserl distinguishes in FTL between manifold theory and ontology proper.[30] All ontology, even 'formal ontology,' aims at the achievement of truth about a region of objects. Manifold theory however introduces 'no cognitional considerations other than those of 'non-contradiction,' of immediate consequence or consistency, which manifestly includes all questions of mathematical 'existence'' (FTL 145 [140]). The thought-objects of manifold theory are unique in that their 'being' consists entirely in their being named in a coherent theory form. They cannot be described except by saying that they are *what* we name when we take the algebraic symbols of such a theory form as names. The objects of a manifold are constituted only when we take the symbols of a theory form as names rather than as mere 'counters' in a 'game' of calculation or deduction.

There is of course a kind of evidence that pertains to manifold theory. In manifold theory as elsewhere, there is a difference between lacking and possessing insight. But Husserl's point is that the evidence achieved in pure manifold theory is in fact the evidence of 'distinctness,' not the evidence of 'clarity.'[31] It is an evidence that pertains to the *meanings* at work, not the *objects meant*. To achieve full understanding of a 'theorem' of non-Euclidean geometry, we need only execute its 'proof' in a thoughtful and explicit manner. We need only return to the 'axioms' of the system and activate their meanings, then derive the theorem in question by a series of actually performed inferences.[32] No further step, no intuitive presentation of the objects meant, is necessary or even possible.[33]

The evidence actually at work in manifold theory is thus not essentially different from that at work in game-mathematics. For there is a difference between vague judging and distinct judging even in the latter; 'even in playing a game one actually judges, once actually collects and counts, one draws actual conclusions, and so forth' (FTL 104 [99]). But why then do we move to manifold theory at all? What leads us to take the symbols in our theory forms as names of thought-objects, if there is in fact no evidence of clarity regarding those objects? We have seen that such symbols do not *have* to be taken as names. We can take them to be meaningful only in the way a rook is meaningful in a game of chess. That is, we can take them to have a meaning that lacks all relation to an object meant. However, it is also possible for us to take them as names of objects, and we do in fact take them as such when we move from deductive games to manifold theory. But there must then be a significant difference between a chess piece and an algebraic sign. For clearly, we cannot take a rook to 'designate' an object; we cannot take a rook to be meaningful in the way *names* are meaningful. What is it then that allows this difference between a chess piece and an algebraic symbol? Why is it that we can take the latter but not the former to name an object?[34]

Although Husserl does not address this question directly in any of his published works, we might find intimations of an answer in his reference in FTL to what he calls 'the sound core of Hermann Hankel's ... 'principle of the permanence of formal laws'' (FTL 101 [97]). This principle is commonly interpreted as a rule governing the extension of mathematical operations; it tells us to choose the extension that leaves the extant calculating rules intact as far as possible.[35] But Husserl might have had a somewhat different interpretation in mind when he referred to the 'sound core' of Hankel's principle. He might have interpreted the principle to refer to a certain 'permanence' of *intentional structures*. That is, he might have found it to have a descriptive, phenomenological sense, in addition to the prescriptive sense ordinarily understood. We noted in the previous section that in universal arithmetic, before the formalizing shift to the theory of theory forms, ' a+ b = b + a' expresses a law to which ideal numbers are necessarily subject. Now in manifold theory, this is obviously no longer the case; the literal symbols are not taken to refer to ideal numbers, nor does '+' designate the addition of numbers (cf. LU I 249 [249]). Nonetheless this law might well be preserved in manifold theory, and preserved precisely in its sense as 'law.' Even though in manifold theory we no longer take 'a + b = b + a' as a law governing the region of ideal numbers, we can continue to take it as a law of *some* kind. In this case, one aspect of the original arithmetical meaning of the law (its reference to the region of numbers) has receded, but another aspect (its law-like character) continues to be at work. In Husserl's terms, a 'sedimented' meaning is still operative, even though the context in which that meaning initially arose has been displaced.[36]

If this interpretation is correct, it allows us to see how Husserl might have dealt with the problem of the difference between a rook and an algebraic symbol. A rook has no meaning prior to its involvement in the game of chess; before becoming a game piece, it was in fact just a bit of 'ivory or wood.' No relevant sedimentation is at work here. But the case is different with the literal symbols employed in the theory of theory forms. These symbols function as parts of larger wholes in which sedimented meanings of the kind indicated are operative. They figure in manifold theory as parts of formal expressions like 'a + b = b + a' – expressions which can retain the sense of being 'laws' even though they are no longer taken to refer to original object regions. But if 'a + b = b + a' is a 'law' and not just a rule of calculation, then clearly there must be something for this law to be 'about.' There must be a region of objects which is subject to the law. The symbols *a* and *b* must therefore be able to work as names. It would be then the sedimented meanings at work in this and other formal laws that lead us to posit the thought-objects of manifolds. It is because of sedimentation that we can and do move beyond game-mathematics to manifold theory.

Husserl's concept of manifolds and manifold theory is, as we can see, extremely subtle. It is only when we thematize the intentional structures involved that we can recognize the difference between manifold theory and game-mathematics. In regard to the actual theory forms developed, there is no difference between them at all. The difference lies entirely in how we 'take' the theory forms. If we take them to define correlative object regions, we are doing manifold theory; if not, we are doing something else.[37]

The difference between manifold theory and game-mathematics is subtle, but it is by no means trivial. Manifold theory provides a mediating link of sorts between game-mathematics and what Husserl calls 'truth-logic.' As Husserl explains in FTL, truth-logic develops when we re-activate the interest in truth that is partially suspended in logical study of the relations between judgments (FTL 60, 70 ff. [55, 65 ff]). We become concerned with the possible truth as well as with the consistency and formal coherence of the apophantic structures that we study. We become concerned with the possible correspondence of these structures to something objective, to something that has 'true being.' Now manifold theory is not itself truth-logic. 'The theory of manifolds [*Mannigfaltigkeiten*] has no compelling reason to include in any manner within its theme questions about the possible actuality (the possible true being) of any single manifolds subsumed under its formal ideas of a manifold' (FTL 143 [138]). What Husserl calls 'the mathematics of the mathematicians' (FTL 146 [141]) – and this obviously includes manifold theory as well as game-mathematics – remains ultimately on the level of apophantics. Despite the fact that manifold theory takes symbols as names of objects, it does not introduce the notion of 'correspondence' or truth (FTL 150 [144]). The objects which it names are available only insofar as they are nameable within the theory form itself. The objects of a manifold are ultimately 'supposed objects as supposed – that is to say, object-senses, as substrate senses, that are adapted to function harmoniously in a judgment-system as substrates of predication' (FTL 148 [142]).[38] It is only when we move beyond manifold theory to truth-logic, Husserl says, that we view mathematics as something that 'relates essentially to *possible* object-provinces and theories' (FTL 146 [140]).[39] But the distinction between manifold theory and truth-logic should not lead us to forget the distinction between manifold theory and game-mathematics.[40] It is only in the latter that the truth interest is *totally* disconnected, that we become *wholly* absorbed in the game of manipulating symbols. Manifold theory does not fully re-activate the truth interest, but the truth interest is at work in manifold theory in a way that it is not at work in game-mathematics. Manifold theory does at least direct itself toward a *formally* possible province of cognition. As such it anticipates the further move to truth-logic, where we ask about *actual* as well as formal possibility.

If there were not a difference between game-mathematics and manifold theory – in other words, if we could not take the symbols used in the study of theory forms as names – formal mathematics would in fact be no different, theoretically speaking, from chess. But formal mathematics *is* different from chess. By introducing the notion of manifold theory and studying the intentional structures at work in it, as Husserl does, we come to see just how it is different.

3. ANALYSIS AS MANIFOLD THEORY

In the preceding section we have discussed manifold theory in general terms. We shall now proceed to see how Husserl used this concept to clarify the sense of mathematical analysis. According to Husserl, analysis is a branch of manifold theory. The 'real number system' whose properties are the central theme of analysis is in fact, as Husserl notes in LU and FTL, a mathematical manifold, a formally defined object region in the sense described.[41] It is for this reason that he writes in FTL that 'a theory of 'real numbers' ... belongs in the formal mathematics of the higher level' (FTL 103 [99]). It is also why he assigns mathematical analysis to the sphere of 'pure apophantic analytics' (FTL 60 [55]). Analysis is not a branch of formal ontology, as arithmetic is.[42] The 'numbers' it studies are not objects generated through eidetic variation over a class of concrete objects. Its numbers have a kind of being, a phenomenological character, that is entirely different from that of the original ideal numbers. The numbers named in analysis are in fact the 'thought-objects' (*Denkobjekte*) of a manifold. They are therefore objects which cannot be presented with the evidence of clarity, objects which have no presence at all apart from the theory forms in which they are named. We can describe these objects only by saying that they are *what* we name when we take the algebraic symbols in certain theory forms as names.

How does the notion of manifold theory help Husserl to clarify the sense of analysis? How does it allow him to explain what we mean by the 'real number system'?

It is sometimes said that negative, rational and irrational numbers have the same kind of being, the same kind of objectivity as the numbers constituted in counting and collecting sensuous groups. We might take Richard Dedekind as a representative of this view, as Husserl himself does in his 1901 *Doppelvortrag*. Dedekind says that all numbers are 'created by the human mind': some are created through the act of counting, others through various acts of definition (PA 434). Negatives, rationals and irrationals arise when we simply decree that there shall be further numbers, alongside those we created first, having certain specified characteristics. On this view there would seem to be no essential difference of ontological status between the original numbers and those introduced in analysis. Now Husserl is quite forceful in expressing his opposition to this view. We cannot simply broaden the original number realm, he claims, by arbitrary decree.

> Once a word – for example, the word 'number' (*Anzahl*) – has been restricted to a given and possibly evident region of objects, I cannot decree by some arbitrary determination or other that the region shall admit a broadening by means of new objects. That would be as if one wanted to decree in geometry that 'there shall be round squares,' if not in the plane, then in a higher dimension of space (PA 435).

What Dedekind's view fails to take into account is the fact that the original numbers are 'given,' not just created. It is true of course that they are not given in the way sensuous objects are given. If we are to make number present, we must perform the appropriate categorial acts; we must actually determine 'how many' items there are in a presented group. But the acts in question do in fact generate a certain presence, and this presence delimits our possibilities of meaning. We can imagine a 'how many' that is fewer than three, fewer than two, even fewer than one. We cannot however imagine a 'how many' that is fewer than none. This latter possibility if 'contradictory to the idea of the how many' (PA 434 f.). So long as number is understood as determinate multitude, or as a mode of 'how many,' 'I cannot without absurdity broaden the sphere of the number concept arbitrarily on the basis of creative definitions ...' (PA 435).

But given this absurdity, how can we allow negatives, rationals and irrationals to have being at all? Husserl explains the matter this way:

> It is certain that we cannot arbitrarily broaden the number concept (*Anzahlbegriff*). But we can indeed abandon the number concept and define a new, purely formal concept by means of the formal system of definitions and operations valid for numbers: the concept of positive whole numbers (*Zahlen*). And this formal concept of positive numbers can be broadened without contradiction by new definitions, just as it is itself delineated by definition (PA 435).

We cannot broaden the original number region without absurdity, but we can *abandon* the region and move to manifold theory (cf. PA 431). That is, we can take arithmetic, the theory of numbers and number relations, as a mere theory form which refers only to an object region which it itself generates. What we call the 'positive whole numbers' (or 'positive integers') are in fact the thought-objects of a manifold generated in this way. Having made the move to manifold theory, the move from *Anzahl* to *Zahl*, we can then without absurdity vary the theory form in question and thus obtain a broader manifold which includes the initial manifold as a special case.

In manifold theory we are not concerned with absurdity or material countersense. 'In this universality, which is not subject to the limitations imposed by given concepts, there is no absurdity (*Widersinn*), unless of course one has provided the definitory operations with contradictory conventions or axioms' (PA 437). Manifold theory is based solely on the evidence of distinctness, as we saw in the preceding section. The evidence of clarity is not at work here. Despite the fact that we take the symbols we use as names of objects, it is nonetheless the distinctive nature of these objects to have no presence apart from the theory form in which they are named. And so long as we are 'abstracting from every effect produced by clarification,' 'material countersense' as distinct from analytic countersense does not become an issue (FTL 70 [65]).

To understand the 'broadening' that leads to negative, rational and irrational numbers, we must be sensitive to the difference between arithmetic taken as a *theory* and arithmetic taken as a *theory form*. When arithmetic is taken as a theory, 'a + b = b + a' expresses an essential, ontological truth about ideal numbers. When arithmetic is taken as a theory form, on the other hand, 'a + b = b + a' functions as part of an axiom system which first defines the object region which it is about. There is a difference between these two ways of taking arithmetic – a difference between ontology and manifold theory – even where the theory form has not yet been altered or expanded. We find something analogous to this in the sphere of geometry. There is a difference between Euclidean geometry 'as Euclid himself understood it' and Euclidean geometry taken as manifold theory (cf. FTL 96 f. [92 f.]; LU I 250 f. [242 f.]). Euclid himself understood geometry as ontology, as 'the theory of intuited world space.' But in the context of manifold theory, what

Euclidean geometry is understood to be 'about' is simply 'Euclidean space' (FTL 97 [94]). That is, what we now intend is simply the 'Euclidean manifold,' a merely possible object region distinguished by being 'three dimensional' and possessing a certain degree of 'curvature' (cf. LU I 250 f. [243]). It becomes possible at this point, of course, to construct other 'geometries' relating to manifolds with different dimensions and curvatures. But it is extremely important to note that we can leave the sphere of ontology even before we begin to make such constructions. We cease to activate the evidence of clarity, to anticipate a presence of the things themselves, even while remaining with the theory form of Euclidean geometry and with the notion of a correlated manifold.[43]

It is only when we take arithmetic as a theory form which defines a manifold that '1' and '0' can be taken as number names. We saw in a previous chapter that there is something number-like about 'one' and 'none,' in the sense that these words can also serve in certain contexts to answer the question 'how many.'[44] Nonetheless one and none are not numbers in the original sense, according to Husserl; they are not *Anzahlen*.[45] What 'one' and 'none' name is not made present in the way what 'two,' 'three' and 'four' name is made present. Different intentional structures are at work in the two cases. But when we move to manifold theory, the difference between 0 and 1, on the one hand, and 2, 3, 4, etc., on the other, begins to disappear.

> None and one are possible results of arithmetical problems, results which come up often enough. Thus an algebraic calculus which seeks to take account of all conceivable cases cannot make a distinction between none and one, on the one hand, and the rest of the numbers on the other. The signs a, b, c ... x, y, z of universal arithmetic must be signs which can be particularized as 2, 3, 4, ..., and 0 and 1 just as well (PA 431).

Even in the 'algebraic calculus,' of course, 0 and 1 have certain distinctive properties. They appear as 'limit cases' of the various operational forms (PA 133). 0 is particularly anomalous, since division by 0, for example, leads to contradictions. 0 cannot therefore be admitted to the region unless axioms are modified or supplemented in some way. Such modifications or supplementations of the original theory form are not necessary in connection with the introduction of 1 to the system. But even the admission of 1 as a 'number' alongside 2, 3, 4, etc., involves a shift of intentional structures which we must not overlook. When we admit 1 as a number, we are no longer taking arithmetic as ontology, but rather as a theory form which defines its own manifold. When we take arithmetic as an actual ontological theory, we are still activating the evidence of clarity. We still anticipate a presence of the numbers we intend. And since a presence of what is 'one' is not as such a presence of number, the distinction between one and number is retained.[46] This distinction vanishes only when we move to manifold theory. Only in the context of manifold theory, where the evidence of clarity is suspended, can arithmetic be taken to be 'about' what are called the positive whole numbers.[47]

Once we have made the move from ontology to manifold theory, we are in a position to make broadenings of the theory form of arithmetic. It is at this point, for example, that what are called 'rational numbers' arise. There are no 'fractions' in the original number sphere. Fractions as such are not determinate multitudes, nor do they arise by supposing that the units of ideal numbers are in some way 'divisible' (PA 436). It is in the context of a theory of what Husserl calls 'divisible magnitude' (*teilbare Grösse*), not in that of the theory of number, that fractions make their appearance (PA 436). Now the theory of divisible magnitude 'leads to an operation-system' just as the theory of number does (PA 436). And once this operation-system has been expressed in suitably general form, it can become apparent that it partially coincides with that of number. The two algorithms 'have quite different spheres, but an analogous structures; they have in part similar operationforms, though different operation-concepts' (PA 436). The difference between the two algorithms can now seem to consist solely in the fact that the one 'includes more basic elements and axioms' than the other (PA 436). And it is at this point that the difference between the two original object regions themselves becomes secondary. The original difference between number and magnitude gives way to the notion of a single object region that is directly correlated with the broader theory form itself: a manifold which we know as the system of rational numbers.

It would of course be absurd to speak of 'rational numbers' if 'number' were still understood in the original sense, just as it would be absurd on this understanding to speak of 0 and 1 as numbers. But in manifold theory we have abandoned the number region strictly speaking, the region of *Anzahlen*. We have moved from ontology to manifold theory, from *Anzahl* to *Zahl*. The objects we now intend have no presence apart from the theory forms that we study. They are, in Husserl's terms, mere 'thought-objects.' And the same is true of negative and irrational numbers, of course. We need not investigate in detail the further broadenings which lead to the introduction of these numbers, and eventually even to the so-called 'imaginary' numbers (cf. LU 250 f. [242 f.]). The intentional structures that make these broadenings possible have already emerged. In admitting such objects, we are not actually adding new objects to the region of number. We are not broadening the concept of number in the way we would broaden our concept of stars should a new star be discovered with properties unknown in the past. We are not even broadening it in the way we would broaden our concept of automobiles should a new kind of automobile be invented. What is distinctive about the 'broadening' that leads to the real number system is that is involves *abandoning* the original object region altogether. Before this broadening can take place, we must cease to intend numbers and begin to intend manifolds instead. We must, in other words, make a shift to the apophantic domain, to the domain of meanings as distinct from objects meant.[48]

We see therefore that analysis is not a science of numbers and their relations, but rather a branch of manifold theory. But we should not overlook the fact that number in the strict sense (*Anzahl*) is still somehow at work in analysis. Analysis is still a way of having numbers in their absence. In Chapter I we noted that Husserl continued to insist on the need for a careful study of the concept of number even after he had recognized that analysis is not a science *of* number.[49] As he puts it in FTL, 'a formal mathematics of the higher level, the manifold-level (*Mannigfaltigkeitsstufe*), can never enable us to do without the specifically logical categories ... and the actual axioms relating to these' (FTL 103 [98 f]). We can now see why this is so. Analysis does not, as it is sometimes thought, wholly absorb the original arithmetic.[50] As Husserl says, 'the arithmetic of cardinal numbers has an existence of its own' (FTL 103 [99]). The numbers studied in arithmetic would exist even if there were no analysis. Although analysis might be said to contain the *theory form* of arithmetic within itself, it cannot generate arithmetical theory as such 'from its own resources' (FTL 103 [99]). On the contrary, analysis seems actually to presuppose arithmetic. That is, it seems to presuppose the intentional structures at work in arithmetic in the original, ontological sense. As manifold theory, analysis is something more than a 'deductive game with symbols.' But it can be this something more, it can lay claim to being a serious cognitive enterprise, only because we can take the symbols at work in it as names.[51] And if the interpretation suggested in the previous section is correct, the fact that algebraic symbols can be taken as names is due ultimately to the 'permanence' of the sedimented meanings at work in them. It is through such sedimented number meanings that analysis, even though it may have lost all orientation toward the presence of number, is nonetheless still related to number. If we could not somehow have numbers in their absence, if numbers were not such as to allow us to have them in absence as well as presence, analysis itself would not be possible in the form in which we know it.

4. HUSSERL'S ATTEMPTED JUSTIFICATION OF ANALYSIS

In the preceding sections of this chapter we have examined Husserl's final answer to a question which he first raised in the Preface to PA: the question as to the 'true sense' of analysis (PA 7). We have seen that he ultimately answered this question by introducing the notion of manifolds and by characterizing analysis as manifold theory. But Husserl also raises a second question in the Preface to PA. This second question concerns the 'justification' (*Rechtfertigung*) for the introduction of negative, rational, irrational and also imaginary numbers (PA 7). Here, it seems, what is at issue is not just the sense, but the very legitimacy of the discipline we know as analysis. Let us see how Husserl understood and dealt with this problem.

Husserl's statement of the problem in PA is brief but revealing. He writes:

> The logical investigation of the arithmetical algorithm – still in its interpretation as an arithmetic of number (*Anzahl*) – and the justification for the calculative use of the quasi-numbers that arise from the inverse operations (the negative, imaginary, fractional and irrational numbers) will be contained in the first part of the second volume.[52]

Now it is important to take note of a certain imprecision in this statement. Husserl seems not to take account of the difference between a manifold and an original object region when he suggests that the 'quasi-numbers' in question are introduced into 'the arithmetical algorithm – still in its interpretation as an arithmetic of number.' As we have seen in the previous section, it is Husserl's mature view that we do not in fact introduce these numbers into 'arithmetic' in the sense of a science of numbers and their relations. When we introduce them, we have already abandoned the original number region and moved to manifold theory; we are already intending the 'positive integers' or 'positive whole numbers,' not numbers in the sense of *Anzahlen*. And from this point of view, it seems somewhat inaccurate to single out the negatives, rationals, irrationals and imaginaries as 'quasi-numbers.' For in fact, even the positive integers themselves could be called 'quasi-numbers.' They too are understood only as the correlates of a certain theory form, not as ideal objects capable of being presented with the evidence of clarity. Now we shall see below that, at a later stage in his career, Husserl came to ask in effect about the 'justification' for the move from ontology to manifold theory, or the move from numbers in the sense of *Anzahlen* to numbers in the sense of positive integers. This is not the question which he asks in PA, however. At this time his question concerns only the justification for a move made *within* manifold theory, so to speak. He is concerned only with the move from the positive integers to the real number system (and eventually with the further move to the complex number system).

Although Husserl never actually published the results of his study of this problem, he evidently worked on it well into the 1890's.[53] And it seems that in his own view at least, he eventually solved the problem. In LU he refers to 'the only possible solution of the problem that has not yet been cleared up: how, e.g., in the field of numbers impossible (essenceless) concepts can be treated like real ones' (LU I 250 [242]). He does not explain this solution in LU itself, but apparently he did do so in his 1901 *Doppelvortrag* on the philosophy of mathematics. From the existing manuscript of this lecture,[54] together with references to the lecture in some of Husserl's later writings,[55] we can form a fairly clear idea of just what his solution was.

We should note first of all that, in the lecture manuscript, Husserl removes the imprecision which we found in his earlier statement of the problem. He now states the problem in such a way as to take account of the difference between a manifold and an original object region. To be sure, he does give an initial formulation which seems *not* to acknowledge this distinction. He observes at first that where a region of objects is given, certain combination forms will have 'no real meaning' (PA 433). 'On the basis of the special nature of the objects' in question, these combination forms will be 'absurd' (*widersinnig*), Husserl says (PA 433). Given the original number region, for example, the concept of a negative, rational or irrational number is absurd or 'imaginary' (PA 432 f.).[56] It is against this background that Husserl first states what he calls 'the problem of the imaginary':

> With what right may something absurd be processed calculatively? With what right can something absurd be applied in deductive thinking as though it were something coherent? How are we to explain that one can operate with something absurd according to rules, and that if the absurdity is dropped from the propositions, the resulting propositions are correct (PA 433)?

But this initial formulation is soon replaced by another one that is more precise. We are not really operating with something 'absurd' when we introduce negative, rational and irrational numbers, for even before introducing them 'one abandons the original object region ... and brings into view, with the most universal universality, a region-in-general that is defined by such operations forms' (PA 437). In other words, one moves from ontology to manifold theory. And on this level of formal universality, as Husserl notes, there is no genuine 'absurdity,' for we are no longer referring to objects that could be given intuitively (PA 437). When we introduce the negatives, rationals and irrationals, we are merely moving from one manifold to another. Properly understood, therefore, 'the problem of the imaginary' concerns only the legitimacy of a move made within manifold theory, the move from the 'narrower' manifold of positive integers to the 'broader' manifold of the real number system.[57]

Now it might seem that we would be justified in making such a move whenever it could be shown that the axiom system defining the broader manifold is consistent. But Husserl claims this is not sufficient. We are justified in broadening our manifold, he contends, only if the axiom system in question has a further property as well. In his view, the crucial question is the following:

> How do we know that what is free of contradiction is also true; or, as it is to be expressed here, how do we know that a proposition which consists exclusively of concepts arising in the narrower [system[58]] and defined there, and which does not contradict the axioms of the narrower region – how do we know that such a proposition holds for the narrower region? ... How do we know, that is, that a proposition which does not contradict the narrower axioms is a consequence of these axioms (PA 439)?

This somewhat abstract formulation of the problem can be expressed more concretely with reference to a particular case. Suppose the narrower manifold in question to be the system of positive integers, the broader manifold to be that of *all* integers, positive and negative alike. When we move from the former system to the latter, we adopt a new set of axioms which allows certain operations previously prohibited (in the broader system, $a - b = c$ even where $a < b$). But we will still be able to formulate and derive propositions in the broader system which are about the positive integers in particular; in other words, we will still have propositions that would be 'meaningful' even in the original system. Now Husserl's question concerns the source of our assurance that such propositions are in fact 'true' within the narrower, original system; or more precisely, our assurance that such propositions follow from the axioms established for the narrower system. How can we be sure, he asks, that a proposition which refers only to the positive integers, but is derived from axioms that admit negative integers as well, is in fact a consequence of the axioms established for the positive integers alone?

In the existing manuscript of the 1901 lecture, Husserl does not formulate a clear answer to the question he poses.[59] In FTL however the answer is stated as follows: 'If the systems are 'definite,' then calculating with imaginary concepts can never lead to contradictions' (FTL 101 [97]). Now this 'answer' is somewhat puzzling, for the question was not just whether the derived propositions would *contradict* others in the system, but rather whether they would actually be *consequences of* the axioms of the original, narrower system. For contradictions to arise, the broader system would have to be an inconsistent one; but we assumed it to be consistent in our formulation of the problem (cf. PA 439). Presumably, then, what Husserl means to say is that the propositions in question will in fact be consequences of the narrower axioms, provided only that both it and the broader system have the property of being 'definite.' But what is a 'definite' axiom system? A 'definite' system is one 'distinguished by the circumstance that any proposition ... that can be constructed, in accordance with the grammar of pure logic, out of the concepts ... occurring in that system, is either 'true' – that is to say: an analytic (purely deducible) consequence of the axioms – or 'false' – that is to say: an analytic contradiction –; *tertium non datur*' (FTL 100 [96]). In other words, as Husserl himself acknowledges, what he means by the 'definiteness' of an axiom system is what Hilbert and others mean by 'completeness.'[60] Husserl's answer to the question posed could therefore be formulated as follows: any consequence derivable within a broader axiom system but formulable within a narrower one will be a theorem of the narrower system whenever both systems are complete. For example, a proposition that refers only to positive integers will be a theorem in the axiom system of positive integers if it can be deduced within the broader axiom system of integers in general, provided that both systems are complete.

What can be said about Husserl's proposed solution to the 'problem of the imaginary'? We might note first of all that if any two such systems were in fact known to be complete, Husserl's answer to the question posed would be a satisfactory one. All that is needed, in fact, is for the narrower system (in our example, the system of positive integers) to be complete, not for both systems to be complete. For, assuming the broader system to be consistent, no proposition derivable within this system could contradict an axiom of the narrower system. And thus any proposition formulable within the narrower system would have to be a theorem of the narrower system, provided only that the narrower system were complete. But in fact, of course, as Goedel proved in 1931, no axiom system at this level of complexity is both consistent and complete. Goedel showed that 'there is an endless number of true arithmetical statements which cannot be deduced from any given set of axioms by a closed set of rules of inference.'[61] The system of positive integers is not complete , and thus we have no assurance that a proposition formulable within the system of positive integers and known not to contradict the axioms of this system will actually be a consequence of these axioms. Thus it is clear, given Goedel's proof, that 'the problem of the imaginary' cannot be resolved in the way Husserl believed it could be.[62]

But is the 'problem of the imaginary,' as Husserl formulates it, really a crucial problem at all? Is it really essential to the philosophy of mathematics that a solution to this problem be found? It is by no means clear that Husserl himself continued to regard it as such in later years. When he refers in *Ideen* and FTL to his earlier work on this problem, his remarks seem to be primarily autobiographical in character.[63] In neither case is it necessary to interpret what he says as implying continued approval of his earlier work. Husserl does of course continue in these later writings to attach importance to the concept of completeness or definiteness. But in *Ideen* and FTL his only claim is that this concept expresses an 'ideal' which mathematicians have in fact been pursuing in ontology and manifold theory alike.[64]

But even if we are correct in thinking that the later Husserl had ceased to regard the 'problem of the imaginary' as a crucial philosophical issue, this does not mean that he had abandoned the task of justifying analysis altogether. In his earlier work he had attempted to justify a move made *within* manifold theory, the move from the narrower manifold of positive integers to the broader manifold of real numbers. In FTL however he seems to suggest that what needs philosophical justification is the move to manifold theory itself. As he puts it, 'the *logician* ... will not easily come upon the thought of making this reduction to an analytics of pure senses' (FTL 145 f. [140]). By what right, Husserl seems to be asking, do we move from ontology to manifold theory? By what right do we exclude the evidence of clarity in favor of the evidence of distinctness? By what right do we move from *Anzahlen* to mere *Zahlen*? These questions are obviously of a more fundamental character than any that might be raised about moves made within manifold theory itself.

To say that the logician will not 'easily' come upon the thought of manifold theory is not to say that he will not come upon it at all. Indeed, Husserl himself claims in the *Krisis* that 'the process whereby material mathematics is put into formal-logical form, where expanded formal logic is made self-sufficient as pure analysis or theory of manifolds, is perfectly *legitimate*, even necessary ...' (*Krisis* 46 [47]). But Husserl's remarks in FTL do suggest that the justification for the move from arithmetic to analysis is not self-evident. Apophantic mathematics cannot be justified in the way ontological mathematics can be. Husserl requires us to think explicitly about the theoretical justification for the former. Just why is it that 'the logician' in Husserl's sense, the man who is interested in truth as well as in consistency and coherence, will want to appropriate apophantic mathematics? What need does the logician have for formal systems developed under partial suspension of the truth interest? These questions, though of great importance, would lead us far beyond the confines of the present study.

NOTES

1 See above, pp. 1 ff.
2 Cf. Husserl, 'Entwurf einer 'Vorrede,'' pp. 126-128.
3 Ms. K I 28/25b.
4 Husserl calls particular attention to '3 grosse Erfindungen, stufenweise aufeinander folgend und aufeinander fussend ...: 1) die Erfindung der algebraischen Zeichensprache ..., 2) die Erfindung der Koordinatenmethode in der Geometrie durch Descartes ..., 3) die Erfindung der Differential- und Integralrechnung durch Newton und Leibniz ...' (Ms. K I 28/25b).
5 '... Theorien und Methoden erstanden, von einer Allgemeinheit und Tragweite, welche der mathematische Geist der Griechen niemals geahnt und erstrebt hat' (Ms. K I 28/25b).
6 Husserl claims in FTL that, as a matter of fact, the Greeks did not yet possess a purely formal concept of number (FTL 80 [84]). He does not however say that they could not possibly have possessed such a concept; and it is in fact difficult to see how Plato's concept of mathematical number, for example, could be described as non-formal.
7 *Ideen* 31 (64); cf. LU II/1 110 f. (341).
8 'Universal arithmetic' in this sense should not be confused with analysis as such. Husserl himself seems not to have distinguished between these early in his career (cf. PA 294). Later, however, he did so (LU I 172 [181]) with LU I 252 f. [243 f.]).
9 Such general solutions become possible, of course, only where there is a clear distinction between symbols for 'unknowns' and symbols for 'parameters.' It was Vieta's achievement to recognize the need for such a distinction; cf. Boyer, *History of Mathematics*, p. 34 f.
10 Husserl says that Vieta's establishment of algebra allowed a 'reduction of the theory of numbers and quantities to a deductive technique' (FTL 84 [80]); cf. PA 432).
11 Ms. K I 25/31a-32a. The exact date of this manuscript is uncertain, but it appears to stem from the early 1890's.
12 Ms. K I 25/31a. Cf. Ms. F I 16/10a: 'Die ursprünglichen mathematischen Disziplinen der rein logischen Sphäre gehen aus von gegebenen rein logischen Grundbegriffen und von Axiomen in echtem Sinn, von unmittelbaren Einsichten, Gesetzen, die im Wesen dieser rein logischen Kategorien gründen. Z.B. der Begriff der Anzahl ist seinem Wesen nach gegeben, und zu ihm gehören die primitiven Zahlgesetze, die gegebene unmittelbar einsichtige Wahrheiten sind. Darauf bauen sich die abhängigen Zahlgesetze und so konstituiert sich wirklich eine Wissenschaft über ein vorgegebenes und bestimmtes, wenn auch kategoriales [?] Gebiet.'
13 Ms. K I 25/31a.
14 Cf. LU/1 328 (518): 'The 'direct operations' $a + b$, ab, a^b yield resultant numbers unrestrictedly, the 'inverse operations,' $a - b$, a/b, $\sqrt[b]{a}$, ${}^b\log a$, only in certain conditions.'
15 Husserl suggests in one early manuscript that we could in fact register such truths, and thus develop a universal arithmetic, even without the use of literal symbols (PA 378).
16 Ms. F I 16/6a: 'Diese Form tritt in Reinheit hervor erst in der mathematischen Methode: nämlich durch die algebraischen Beziehungen und die sinnlich äusserlichen Formen der Zusammenhänge ... Danach kann es ganz gut kommen und es kommt oft genug vor, dass zwei in ganz verschiedene Sachgebieten etablierte mathematische Theorien in der Form völlig übereinstimmen, weil beide von Grundsätzen ausgehen und Grundbegriffen, die inhaltlich zwar verschiedene Bedeutung haben, aber formal von völlig gleicher Konstitution sind. Bedeuten z.B. die Buchstaben Mengen oder Zahlen und + die Verknupfung von Mengen wieder zu Mengen, von Zahlen wieder zu Zahlen, ... so gilt der Satz $a + b = b + a$. Wir können aber unter den Buchstaben auch Strecken verstehen und unter dem Pluszeichen die Heranschiebung der einen Strecke an die andere ... Dann gilt wieder $a + b = b + a$. Der Satz hat zwar einen anderen Sinn, ... aber der Form nach stimmt dieser Satz ganz mit dem vorigen überein und erhält daher auch denselben algebraischen Ausdruck.'
17 Ms. K I 16/16b: 'Nach dieser Beobachtung ist es also klar, dass man die Form des mathematischen Systems von seinem Gebiet, gleichsam seiner Materie emanzipieren ... kann.'
18 Ms. K I 16/16b: 'Natürlich braucht man die Deduktionen also nicht zweimal wirklich

auszuführen.' It is interesting to compare this comment with Aristotle's remarks on the general theory of proportions of Eudoxus: 'Just as it used to be proved separately that a proportion can be alternated insofar as it is a proportion of numbers and lines and solids and times, so now it can be shown for all in one demonstration; ... they were taken separately, but now it is proved generally' (*Posterior Analytics* 74 a 17ff; cf. Klein, *Greek Mathematical Thought*, p. 158 f.). But it seems that the ancient notion of a *general theory* (or 'general treatment,' *katholou pragmateia*) must be distinguished from what Husserl calls a theory form. In developing a general theory, we intend what is common to things of several different kinds; as Aristotle says, 'the general propositions in mathematics are not about separate things which exist outside of and alongside the magnitudes and numbers, but are just about these ...' (*Metaphysics* 1077 b 17-19). In studying a theory form, however, we cease to intend the original 'things themselves' altogether. Ancient mathematics never took the step which Husserl describes as the 'liberation' of the theory form from the original object regions.

19 FTL 93 (90); cf. LU I 240 (248).

20 The distinction in question is not equivalent to that between 'consequence logic' and 'truth logic,' a distinction which *is* applicable to the lower level of apophantic logic (cf. FTL 71 ff. [66 ff.]). On the relation between the theory of theory forms and truth logic, see below, p. 119 f.

21 Cf. Ms. K I 25/33a-33b: 'Betrachte ich in dieser Weise die Zeichen für sich, so sind sie darum nicht blosse Schnörkel auf dem Papier. Sie haben offenbar eine gewisse Bedeutung. Was ist nun ihre Bedeutung? Nicht mehr die entsprechende arithmetische Bedeutung. Denn von dieser habe ich ja gänzlich abstrahiert. Offenbar liegt die Bedeutung jetz in den Spielregeln. Es ist ganz so wie im Schachspiel. Läufer, Türme etc.'

22 PA 258. Willard misinterprets this passage when he writes that 'the division of 1236 by 12 – utilizing, in the common style, the *numerals* (not, of course, the numbers) in the familiar decadal system – provides an example of calculation in this sense'; 'Husserl on a Logic That Failed,' p. 26. Husserl says explicitly that calculation in the sense in question arises from 'higher logical interests than those of the arithmetica numerosa' (PA 258). It is only in the context of the theory of theory forms that this type of calculation, as distinct from the everyday calculation which Willard mentions, can emerge. On the different senses of calculation which Husserl distinguishes, see above, p. 108 n. 38.

23 In translating '*Mannigfaltigkeit*' as 'manifold,' we follow Findlay, Carr and Boyce-Gibson rather than Cairns. The latter uses 'multiplicity' for this important Husserlian term. In citing Cairns' translations of passages in FTL in which this term occurs, we shall change 'multiplicity' to 'manifold' in order to make the correlation with Husserl's other writings more apparent.

24 FTL 97 (93). Cf. S. Bachelard, *A Study of Husserl's 'Formal and Transcendental Logic*,' tr. by L.E. Embree (Evanston, Ill.,: Northwestern University Press, 1968), p. 47. The point is made very clearly in a 1907 manuscript in which Husserl remarks: 'der Name Mannigfaltigkeitslehre hat seine Bedenken, da darunter immer auch die Mengenlehre mit befaßt wird, die nach unserer sorgsamen [?] Orientierung in die andere Schicht gehört' (MS F I 16/3a). Far from being two terms for the same thing, 'set theory' and 'manifold theory' belong to entirely different levels of logic.

25 It was apparently Riemann, not Cantor, who inspired Husserl's use of the term 'manifold'; cf. Bachelard, *Study of Husserl's 'Logic'*, p. 55.

26 LU II/1 249 (241). Husserl quotes this passage verbatim in FTL with the comment that he 'cannot improve on it' (FTL 94 f. [91]).

27 At one point in FTL Husserl says that '*theory forms*, according to their own sense, have as their correlates *objective manifolds*' (FTL 110 [105]).

28 Ms. F I 16/10a: 'Das, was man aber Axiome in der 'nichteuklidischen Geometrie' nennt, das sind in Wahrheit keine vollen Sätze, wie denn die Grundbegriffe in Wahrheit keine Begriffe sind. Und so ist die nichteuklidische Geometrie eigentlich keine mathematische Wissenschaft im gewöhnlichen Sinn, kein theoretisches System von Sätzen, die aus Grundwahrheiten rein deduktiv gründen.'

29 It seems to be Vieta who set in motion the shift from numbers in the original sense (numbers in the sense of *Anzahlen* or *arithmoi*) to numbers in the sense of manifold objects. See Klein, *Greek Mathematical Thought*, pp. 174 ff. According to Klein, the rules governing the manipulation of Vieta's algebraic letters 'represent the first modern *axiom system*; they create the systematic context which originally 'defines' the object to which they apply' (ibid., p. 176). And since the object is defined in this way, its 'being' is to be understood 'neither as independent in the Pythagorean and Platonic sense nor as attained by 'abstraction' (*ex aphaireseos*), i.e. as 'reduced' in the Aristotelian sense, but as *symbolic*' (ibid., p. 175). On Klein's view, Vieta does not actually complete the shift from the science of numbers to manifold theory. Klein claims that it is only the number 'species,' not numbers themselves, that are transformed by Vieta into 'symbolic formations,' 'formations whose merely potential objectivity is understood as an actual objectivity' (ibid.). 'In Vieta's notion of 'species' the original understanding of number is retained, as it is, of course, in the *Arithmetic* of Diophantus. But his immediate successors, Ghetaldi, Harriot, Oughtred and Wallis ... have already lost the original intuition. The technique of operating with symbols replaces the science of numbers' (Klein, 'Phenomenology and the History of Science,' p. 160). Although Husserl's own analyses move on the level of *a priori* possibility, Klein's work shows how fruitful these analyses can be when the categories they generate are used in studying the actual history of mathematical thought. On Klein's relation to Husserl, cf. ibid., p. 143.
30 At one point in FTL, Husserl suggests that manifold theory is 'non-apophantic,' that it belongs with formal ontological disciplines strictly speaking (FTL 81 f. [77 f.]). This initial suggestion is corrected, however, at a later stage in the work (FTL 142 ff. [137 ff]); cf. Sokolowski, *Husserlian Meditations*, p. 285 f.
31 On the distinction between these two kinds of evidence, see FTL 61 ff (56 ff.); also Sokolowski, *Husserlian Meditations*, pp. 210 ff.
32 Cf. Sokolowski, *Husserlian Meditations*, p. 38.
33 Although the attempt is sometimes made to construct intuitive models of certain non-Euclidean manifolds, the evidence on which non-Euclidean geometry as such is founded does not derive from the inspection of these models. This may be what Husserl has in mind when he writes that 'a trace of impurity may remain, since the mathematically constructed forms of manifolds are, as a rule, thought of as forms of possible actualities: provided only that ... this thought never exercises any function in mathematics itself' (FTL 145 [140]).
34 The same question could of course be raised regarding the terms employed in non-Euclidean geometry.
35 F. Waismann, *Introduction to Mathematical Thinking*, tr. by T.J. Benac (New York: Frederick Ungar Publishing Co., 1951), p. 27. Hankel's principle is exemplified, Waisman suggests (ibid., p. 26 f.), in the decision to make $a^0 = 1$. This leaves intact the general rule that $a^m \times a^n = a^{m+n}$.
36 On the notion of 'sedimentation,' see above, pp. 79 ff. The interpretation offered here of Husserl's reference to the principle of permanence is suggested by the following passage in an early manuscript: 'Prinzip der Permanenz. Wenn vermöge der Besonderheit der einen Algorithmus begründenden Begriffe gewisse der algorithmischen Operationen nicht in voller Allgemeinheit ausführbar sind, ohne daß man auf widerstreitende Begriffsbildungen kommt, so erweitert man den Algorithmus, nachdem man ihn von der begrifflichen Grundlage losgelöst und als einen konventionellen gedacht hat, dadurch, dass man jede solche Bildung versuchsweise dem algorithmischen Gebiete adjungiert *und die Konvention hinzufügt, dass auch für die durch sie symbolisierten Gegenstände (Zeichen) die alten Gesetze gültig bleiben* ...' (Ms. K I 36/38; emphasis added).
37 The distinction between manifold theory and game-mathematics is one that becomes important only when we seek to clarify the mathematician's achievement philosophically. As Husserl notes in FTL, the mathematician as such lives 'entirely with a view to discovering new theoretical results,' and thus 'he is not in the least interested in changes that convert an equivalent into an equivalent' (FTL 153 [147]).

38 On Husserl's notion of 'the supposed as supposed,' see Sokolowski, *Husserlian Meditations*, pp. 46 f., 278 ff.
39 Husserl seems at first to contradict himself here, for he elsewhere characterizes a 'manifold' as a 'possible province of cognition' (FTL 94 [91]). This latter statement implies that all manifold theory is concerned with possible object regions. But the appearance of contradiction is removed if we bear in mind the distinction between *formal* and *actual* possibility; manifold theory as such is concerned only with the former, truth logic with the latter as well.
40 In Chapter 3 of FTL, Husserl stresses the distinction between manifold theory and game-mathematics (see esp. FTL 102 ff. [98 ff]). In Chapter 5, he stresses the distinction between manifold theory and truth logic (FTL 142 ff. [137 ff.]). Both distinctions must be borne in mind if we are to understand the specific character of the intentionality at work in the manifold theory.
41 LU I 250, 252 (242, 243); FTL 81 (77).
42 Although Husserl suggests at FTL 81 (77) that 'analysis' is an example of non-apophantic mathematics, this suggestion is later corrected; see above, p. 266 n.
43 It is the difference between ontology and manifold theory that leads Husserl to distinguish between 'space' and the 'categorial form of space' (LU I 251 [243]); FTL 97 n. [94 n.]). 'Space' itself – what Euclid understood his geometry to be about – is an ontological concept. The 'categorial form of space,' on the other hand, is a concept from manifold theory; it refers to the manifold which Euclidean geometry is understood to be 'about' in the context of non-Euclidean geometry. Although we cannot here discuss Husserl's philosophy of geometry in detail, it should be noted that his distinction between geometry as ontology and geometry as manifold theory is by no means obliterated by the applications which have been made of non-Euclidean geometry in connection with the general theory of relativity.
44 See above, p. 67.
45 G. Martin seems to miss this important point when he makes the following interpretive comment regarding PA: 'Im Unterschied zu dieser symbolischen Konstruktion können die Anfangszahlen, unter ihnen vorzüglich die Eins und die Zwei, in direkter Anschauung aufgewiesen werden ...' (*Klassische Ontologie der Zahl*, p. 143). It is only in the context of what Husserl would later call manifold theory, where the intuitive presence of number is no longer at work, that 'one' can become a 'number' alongside two (cf. PA 131 ff.).
46 Husserl notes in passing that 'the Pythagoreans did not yet consider one to be a number' (PA 133 n.). But actually this was true of Greek mathematics as a whole; cf. Klein, *Greek Mathetmatical Thought*, pp. 49, 51 ff. The arithmetic and even the 'logistic' of the ancients never lost its orientation to the presence of number.
47 This suspension of the evidence of clarity is even more obvious where arithmetic is understood to be about the 'natural number series' 0, 1, 2, ... To include 0 as a 'number,' we must actually modify or supplement the theory form of arithmetic.
48 G. Martin raises the question as to how 'real number' can on Husserl's view constitute a 'genus'; *Klassische Ontologie der Zahl*, p. 145 f. As Martin points out, it is certainly not through any '*intuitive representation* of the essence in adequate Ideation' (cf. LU I 244 [238]) that 'real number' comes to be given. Martin contends that the interpretation of real number as a genus, though suggested by Husserl himself (cf. LU 251 f. [243]), is in fact 'not easily unifiable' with the general notion of genus and species. This would seem to be correct. The 'real number system' derives its unity only from its correlation to a unified theory form, not from the intuitive presence of the objects themselves.
49 See above, p. 18.
50 Cf. Ms. F I 16/11a: 'Selbst bei den Mathematikern finden wir, und es ist das eine ganz gewöhnliche Erfahrung [?], die Ansicht, daß diese axiomatische Mathematik alle Mathematik in sich aufsauge.'
51 Klein alludes obliquely to the importance of our ability to take algebraic symbols as names of numbers when he contrasts medieval algebra with what in the modern period came to be known as analysis: 'While ... the 'algebra' which has Arabic sources is continually elaborated in

respect to techniques of calculation, for instance by the introduction of 'negative,' 'irrational,' and even so-called 'imaginary' magnitudes ..., its self-understanding fails to keep pace with these technical advances. This algebraic school becomes conscious of its own 'scientific' character and of the novelty of its 'number' concept only at the moment of direct contact with the corresponding Greek science, i.e., with the *Arithmetic* of Diophantus' (*Greek Mathematical Thought*, p. 148 f.). Before Vieta, algebra appeared to be only an 'art,' not a 'science'; its symbols were not understood to have reference to a corresponding region of objects. After Vieta, however, this was no longer the case. Through his reinterpretation of Diophantus, and thus of Greek 'ontological' arithmetic or logistic, Vieta was able to bring about a 'transformation of the ancient concept of *arithmos*' and thereby 'assign to 'algebra,' to this 'ars magna,' *a fundamental place in the system of knowledge in general*' (ibid., pp. 185, 184). Algebraic symbols had come to be taken as a way of referring to a corresponding region of objects, to what Husserl calls a 'manifold.' It is because Vieta made this possible that Klein regards him as 'the true founder of modern mathematics' (ibid., p. 5).

52 PA 7. The second volume was, of course, never published.

53 See above, p. 15.

54 PA 430-451; see above, p. 16.

55 *Ideen* 168 n. (188 n.); FTL 101 f. (97).

56 Husserl is obviously using the term 'imaginary' in a broader sense than is customary in mathematics. What he here calls 'imaginary' numbers are what he had in the Preface to PA called 'quasi-numbers.'

57 Cf. the formulation of the problem in FTL: 'How far does the possibility extend of 'enlarging' (*erweitern*) a 'manifold,' a well-defined deductive system, to make a new one that contains the old one as a 'part'?' (FTL 101 [97]).

58 The manuscript has '*Begriff*' rather than '*System*.'

59 Husserl does say that 'the passage through the imaginary is bound up with the condition of definiteness ...' (PA 443). It would seem that this point is explained by the following remarks: '... the axiom system of arithmetic ... has the property that every equation falling in this arithmetic either holds on the basis of the axioms or does not hold on the basis of the axioms; i.e. the proposition is either a consequence of the axioms or it is in contradiction with the axioms ... An equation derived in the broader region cannot contradict the axioms of the narrower region; otherwise the whole broader region would be inconsistent. Thus it is correct' (PA 441). Having made these remarks, however, Husserl proceeds to say that they are not sufficient to solve the problem (PA 441). The manuscript breaks off without a clear resolution of the issue.

60 *Ideen* 168 n. (188 n.); FTL 101 (96 f.). The 'definiteness' or 'completeness' of an axiom system in Husserl's sense should not however be confounded with Hilbert's 'axiom of completeness'; cf. Bachelard, *Study of Husserl's 'Logic'*, pp. 59 ff.

61 E. Nagel and J.R. Newman, *Goedel's Proof* (New York: New York University Press, 1968), p. 98.

62 Although Goedel's paper was published in 1931, there is no evidence that Husserl ever took notice of it; cf. Eley, 'Einleitung des Herausgebers,' PA XV n.

63 In FTL Husserl refers to 'the imaginary' as 'the concluding theme of my old philosophico-mathematical studies' (FTL 102 [97]); cf. *Ideen* 168 n. [188 n.]).

64 Cf. Bachelard, *Study of Husserl's 'Logic'*, pp. 53 ff. Bachelard may be somewhat hasty however in concluding that the 'ideal of axiomatic grounding carried out once and for all' is in fact a 'Husserlian ideal,' i.e. something which Husserl thought mathematicians should actually strive for. Some of Husserl's remarks suggest that he is merely trying to *interpret* the direction in which mathematics has moved, to bring out the 'hidden origin' of a concept which has 'continually guided mathematics from within' (FTL 99 [95]). From this it would not necessarily follow that Husserl is *proposing* a further pursuit of the 'ideal' in question. His central point may simply have been that this ideal, which is now at work in manifold theory, nonetheless had its origin in ontology; cf. *Ideen* 166 ff (187 f.).

CONCLUSION

In this study we have examined the major themes in Husserl's philosophy of arithmetic. We have seen that what Husserl meant by the 'philosophy of arithmetic' could be better described as the philosophy of formal mathematics, for he was concerned not only with elementary arithmetic, but also with the higher mathematical discipline of analysis, the discipline in which he himself had received his doctorate.

We have also seen that to understand Husserl's work in this area, one must consider his earlier and later writings together. His early work *Philosophie der Arithmetik* (1891) was superseded in some respects by *Logische Untersuchungen* (1900/1901) and *Formale und transzendentale Logik* (1929). But in other respects it remains an indispensable foundation for the philosophy of formal mathematics presented in these later works.

In considering Husserl's investigations of formal mathematics, we must distinguish between two issues: the issue of clarification and the issue of justification. Husserl sought to provide both a clarification and justification of analysis, but he may be judged to have had considerably more success with the former than he had with the latter.

Husserl's attempt to *justify* analysis – or more precisely, to justify the broadening which leads from the system of positive integers to the real number system – was based on the notion of a complete or 'definite' axiom system. But subsequent developments have shown that there are no complete axiom systems of the sort required by Husserl's theory, and therefore his proposed justification must be called into question.

Husserl's attempt to *clarify* analysis, on the other hand, is unaffected by these developments in logical theory. This issue pertains to phenomenology, not to formal logic as such. To clarify the 'sense' of analysis is to show the intentional structures that are at work in this discipline, the intentional structures that are correlated with the real number system. With his many careful distinctions and detailed descriptive analyses, Husserl may be said to have made a valuable contribution to our understanding of these structures and others which they presuppose.

We have suggested that the central theme of Husserl's phenomenological study of formal mathematics is a distinction he makes regarding 'empty' and 'filled' modes of intentionality. As we have put it, the central theme of Husserl's philosophy of arithmetic is that of the presence and absence of number.

This theme can be observed on two different levels. In Chapters II through IV of this study, we have examined it as it unfolds on the level of concrete number experience. Concrete numbers, as 'determinate multitudes,' are to be considered objects of a distinctive kind. They are what Husserl calls 'categorial objects.' Concrete numbers are not presented in sensuous intuition, but rather in sensuously founded, 'categorial' acts of 'counting,' i.e. of explicating and comparing. Through the performance of such acts, a sensuously presented group comes to be given in a fully explicit manner: each item is given in its own right, and so too are the exact relations of equality and inequality which the group bears to other multitudes.

Small numbers are presented in a way large numbers are not; but, we have argued, large numbers can also be made present in a manner that is appropriate to them.

The presence of number must be understood in contrast to several related modes of absence: the absence of a sensuous foundation for counting; the special case of absence in which what was anticipated as 'many' turns out to be only 'one' or even 'none'; the absences which arise even in the presence of a sensuous group when we fail to perform, or fail to carry through to completion, one or both of the requisite categorial acts (explicating, comparing); and finally, the absence that is at work when we use words or signs laden with sedimented number meanings (as in everyday counting and calculating).

The second level on which the theme of presence and absence unfolds is that of mathematical experience proper, which we have examined in Chapters V and VI of our study. The presence that emerges here is that of what Husserl calls 'ideal numbers.' Ideal numbers are to be understood as the essences or species of the corresponding concrete numbers. They are what we refer to when we make judgments about 'the number Four,' for example, and what is given when such judgments are made with insight.

Ideal numbers are made present only through eidetic variation over the corresponding concrete numbers. They are given intuitively only as something identically the same in an imaginatively constructed class of concrete numbers.

The presence of ideal numbers is to be understood in contrast to certain closely related forms of intentionality. The presence of a number sign is not the same as the presence of an ideal number. Neither is the presence of a single concrete number equivalent to the presence of the corresponding ideal number. Even if we take number as an 'empirical universal,' as an identity in a passively generated class of concrete numbers, ideal number as such has not been made present.

The original arithmetic, which is founded on the presence of ideal numbers, can be considered a branch of 'formal ontology': it discloses necessary truths about a region of formal objects. Mathematical analysis, however, involves a move to an entirely different sense of formality. In analysis we move from formal objects to theory forms, or from numerical formality to algebraic formality.

The real number system whose properties are studied in analysis is actually a 'manifold' which has an objective status comparable to that of the spatial manifolds of non-Euclidean geometry. Manifolds, in Husserl's technical sense of the term, are object regions constituted only as the correlates of theory forms. The constitution of a manifold presupposes an intentional shift to what Husserl calls the 'apophantic' domain, the domain of meanings as distinct from objects meant. Since the original number region, the region of ideal numbers, does not presuppose such a shift, the two senses of number must be sharply distinguished from each other.

We have argued that the theme of presence and absence is of great systematic importance in connection with the notion of manifolds. Even though analysis, as a branch of manifold theory, has abandoned the evidence of 'clarity' in favor of the evidence of 'distinctness,' numbers themselves are still somehow at work in analysis. A presence of ideal numbers is no longer anticipated, but it is nonetheless the 'permanence' of sedimented number meanings that allows algebraic symbols to be taken as names of objects. It is only because of this sedimentation that 'thought-objects' as distinct from meanings can be constituted within the apophantic domain itself.

Early in his career, Husserl considered investigations of number such as those discussed and developed in this study to belong to 'psychology.' In fact, however, this view seems to have been incorrect. In studying the presence and absence of number, we are studying what numbers truly are. What is at issue is not just the nature of the human mind, but rather the true being of numbers themselves. By number we mean precisely something that is available to us as identically the same in certain modes of presence and absence. Husserl's investigations of number, even when labeled 'psychology,' may nonetheless be said to lead us beyond psychology insofar as they make this availability thematic.

NOTE ON ABBREVIATIONS

The following abbreviations have been used in this study for works by Husserl (see Bibliography for complete references):

BdZ = 'Über den Begriff der Zahl'

CM = *Cartesianische Meditationen*, translated as *Cartesian Meditations* by Dorion Cairns

EU = *Erfahrung und Urteil*, translated as *Experience and Judgment* by James S. Churchill and Karl Ameriks

FTL = *Formale und transzendentale Logik*, translated as *Formal and Transcendental Logic* by Dorion Cairns

Ideen = *Ideen zu einer reinen Phänomenologie und phänomenologischen Philosophie*, Book One, translated as *Ideas, General Introduction to Pure Phenomenology* by W.R. Boyce-Gibson

LU = *Logische Untersuchungen*, translated as *Logical Investigations* by J.N. Findlay

PA = *Philosophie der Arithmetik*

In citing these works, we give the page number of the German edition immediately following the abbreviated reference. The corresponding page number of the English translation, if any, follows in parentheses. For example, 'FTL 81 (77)' means '*Formale und transzendentale Logik*, p. 81; English translation, p. 77.'

BdZ has bee published as a supplementary text in the *Husserliana* edition of PA (see PA 289-338). Our references to BdZ are to this edition.

In some cases, translations have been modified slightly in the interest of homogeneity. We have indicated these modifications either explicitly in the footnotes or implicitly by supplying the affected German words in parentheses.

BIBLIOGRAPHY

A. Works By Husserl

'Bericht über deutsche Schriften zur Logik in den Jahren 1895-99.' *Archiv für systematische Philosophie* 9 (1903), pp. 113-132, 237-259, 393-408, 523-543; 10 (1904), pp. 101-125.

Cartesian Meditations: An Introduction to Phenomenology. Translated by Dorion Cairns. The Hague: Martinus Nijhoff, 1970.

The Crisis of European Sciences and Transcendental Phenomenology. Translated by David Carr. Evanston, Ill.: Northwestern University Press, 1970.

'Entwurf einer 'Vorrede' zu den *Logischen Untersuchungen* (1913).' Edited by Eugen Fink. *Tijdschrift voor Philosophie* 1 (1939), pp. 106-113, 319-339.

Erfahrung und Urteil. Untersuchungen zur Genealogie der Logik. Edited by Ludwig Landgrebe. Hamburg: Felix Meiner Verlag, 1972.

Experience and Judgment. Investigations in a Genealogy of Logic. Translated by James S. Churchill and Karl Ameriks. Evanston, Ill.: Northwestern University Press, 1973.

Formal and Transcendental Logic. Translated by Dorion Cairns. The Hague: Martinus Nijhoff, 1969.

Husserliana. Edmund Husserl: Gesammelte Werke. The Hague: Martinus Nijhoff, 1950-

Vol. I: *Cartesianische Meditationen und Pariser Vorträge*. Edited by S. Strasser. 1973.

Vol. III: *Ideen zu einer reinen Phänomenologie und phänomenologischen Philosophie*. Book One. Edited by Walter Biemel. 1950.

Vol. VI: *Die Krisis der europäischen Wissenschaften und die transzendentale Phänomenologie. Eine Einleitung in die phänomenologische Philosophie*. Edited by Walter Biemel. 1954.

Vol. XI: *Analysen zur passiven Synthesis (1918-1926)*. Edited by Margot Fleischer. 1966.

Vol. XII: *Philosophie der Arithmetik. Mit ergänzenden Texten (1890-1901)*. Edited by Lothar Eley. 1970.

Vol. XVII: *Formale und transzendentale Logik. Versuch einer Kritik der logischen Vernunft. Mit ergänzenden Texten*. Edited by Paul Janssen. 1974.

Vol. XXII: *Aufsätze und Rezensionen (1890-1910)*. Edited by Bernhard Rang. 1979.

Ideas. General Introduction to Pure Phenomenology. Translated by W.R. Boyce-Gibson. New York: Collier Books, 1972.

Letter to Meinong of 22 May 1891. In *Philosophenbriefe. Aus der wissenschaftlichen Korrespondenz von Alexius Meinong*, pp. 96-100. Edited by Rudolph Kindinger. Graz, Austria: Akademische Druck- und Berlagsanstalt, 1965.

Logical Investigations. Translated by J.N. Findlay. London: Routledge and Kegan Paul, 1970.

Logische Untersuchungen. Three volumes. Fifth edition of Volumes I and II/1; fourth edition of Volume II/2. Tübingen: Max Niemeyer Verlag, 1968.

'Persönliche Aufzeichnungen.' Edited by Walter Biemel. *Philosophy and Phenomenological Research* 16 (1956), pp. 293-302.

'Psychologische Studien zur elementaren Logik.' *Philosophische Monatshefte* 30 (1894), pp. 159-191.

'(Review of) Schröder, Ernst. Vorlesungen über die Algebra der Logik (Exacte Logik).' *Göttingsche gelehrte Anzeigen* (1891), pp. 243-278.

B. Unpublished Manuscripts by Husserl

F I 16. 1906-1907.
K I 6. 1892-1893 (?).
K I 25. No date.
K I 28. 1887-1890.
K I 36. No date.

C. Works About Husserl

Bachelard, Suzanne. *A Study of Husserl's 'Formal and Transcendental Logic'*. Translated by Lester E. Embree. Evanston, Ill.: Northwestern University Press, 1968.

Becker, Oskar. 'The Philosophy of Edmund Husserl.' In *The Phenomenology of Husserl. Selected Critical Readings*, pp. 40-72. Edited and translated by R.O. Elveton. Chicago: Quadrangle Books, 1970.

Biemel, Walter. 'The Decisive Phases in the Development of Husserl's Philosophy.' In *The Phenomenology of Husserl. Selected Critical Readings*, pp. 148-173. Edited and translated by R.O. Elveton. Chicago: Quadrangle Books, 1970.

Bosio, Franco. 'Gli inizi della fenomenologia. La *Filosofia dell'Aritmetica*.' *Aut Aut* 70 (1962), pp. 294-308.

De Boer, Theodore. *The Development of Husserl's Thought*. Translated by Theodore Plantinga. The Hague: Martinus Nijhoff, 1978.

Farber, Marvin. *The Foundation of Phenomenology. Edmund Husserl and the Quest for a Rigorous Science of Philosophy*. Third edition. Albany: State University of New York Press, 1968.

Föllesdal, Dagfinn. 'Husserl und Frege. Ein Beitrag zur Beleuchtung der Entstehung der phänomenologischen Philosophie.' *Avhandlinger utgitt av Det Norske Videnskaps-Akademie i Oslo. Hist.-Filos. Klasse II*. 1958. No. 2.

Frege, Gottlob. 'Rezension von: E.G. Husserl, Philosophie der Arithmetik. I.' In *Kleine Schriften*, pp. 179-192. Edited by Ignacio Angelelli. Darmstadt: Wissenschaftliche Buchgesellschaft, 1967.

Gurwitsch, Aron. *The Field of Consciousness*. Pittsburgh: Duquesne University Press, 1964.

Haddock, Guillermo Ernesto Rosado. 'Edmund Husserls Philosophie der Logik und Mathematic im Lichte der gegenwärtigen Logik und Grundlagenforschung.' Dissertation, Bonn, 1973.

Huertes-Jourda, Jose. 'On the Threshhold of Phenomenology. A Study in Edmund Husserl's *Philosophie der Arithmetik*.' Typewritten.

Kelkel, Lothar and Scherer, René. *Husserl. Sa vie, son oeuvre. Avec un exposé de sa philosophie*. Paris: Presses Universitaires de France, 1964.
Klein, Jacob. 'Phenomenology and the History of Science.' In *Philosophical Studies in Memory of Edmund Husserl*, pp. 143-163. Edited by Marvin Farber. New York: Greenwood Press, 1968.
Landgrebe, Ludwig. 'Husserls Phänomenologie und die Motive zu ihrer Umbildung.' In *Der Weg der Phänomenologie. Das Problem einer ursprünglichen Erfahrung*. Gutersloh: Verlagshaus Gerd Mohn, 1967.
Mohanty, J.N. 'Husserl and Frege: A New Look at their Relationship.' *Research in Phenomenology* 4 (1974), pp. 51-62.
Picker, Bernold. 'Die Bedeutung der Mathematik für die Philosophie Edmund Husserls.' *Philosophia naturalis* 7 (1961-1962), pp. 266-355.
Pietersma, Henry. 'Husserl and Frege.' *Archiv für Geschichte der Philosophie* 49 (1967), pp. 298-323.
Pivcevic, Edo. *Husserl and Phenomenology*. London: Hutchinson & Co., 1970.
Schuhmann, Karl. *Husserl-Chronik. Denk- und Lebensweg Edmund Husserls*. The Hague: Martinus Nijhoff, 1977.
Sokolowski, Robert. *The Formation of Husserl's Concept of Constitution*. The Hague: Martinus Nijhoff, 1970.
– *Husserlian Meditations. How Words Present Things*. Evanston, Ill.: Northwestern University Press, 1974.
– 'Ontological Possibilities in Phenomenology: The Dyad and the One.' *Review of Metaphysics* 29 (1974), pp. 691-701.
– *Presence and Absence. A Philosophical Investigation of Language and Being*. Bloomington, Ind.: Indiana University Press, 1978.
– 'The Structure and Content of Husserl's *Logical Investigations.*' *Inquiry* 14 (1971), pp. 318-347.
Spiegelberg, Herbert. *The Phenomenological Movement. A Historical Introduction*. Two volumes. The Hague: Martinus Nijhoff, 1971. Third revised edition, one volume, 1982.
Torretti, Roberto, 'La filosofía de la aritmética de Husserl.' *Studi Internazionali di Filosofia* 4 (1972), pp. 183-206.
Tugendhat, Ernst. *Der Wahrheitsbegriff bei Husserl und Heidegger*. Berlin: Walter de Gruyter, 1967.
Verdenal, René. 'La sémiotique de Husserl: la logique des signes. (A propos de certaines inédits).' *Les Etudes Philosophiques* 4 (1973), pp. 553-564.
Willard, Dallas. 'Husserl on a Logic That Failed.' Paper presented at the Eastern Division meeting of the American Philosophical Association, Washington, D.C., December 1978.
Winance, Eleuthere. 'Logique, mathématique et ontologie comme 'mathesis universalis' chez Edmund Husserl.' *Revue thomiste* 66 (1966), pp. 410-434.

D. Other Works

Benacerraf, Paul and Putnam, Hilary, editors. *The Philosophy of Mathematics. A Book of Readings*. Englewood Cliffs, N.J.: Prentice-Hall, 1964.
Beth, Evert W. *The Foundations of Mathematics*. Second edition. Amsterdam: North Holland Publishing Co., 1968.
Boyer, Carl B. *The History of the Calculus and its Conceptual Development*. New York: Dover Publications, 1959.
– *A History of Mathematics*. New York: John Wiley & Sons, 1968.

Brentano, Franz. *Die Lehre vom richtigen Urteil*. Edited by Franziska Mayer-Hillebrand. Bern: Francke Verlag, 1956.
– *Psychologie vom empirischen Standpunkte*. Edited by Oskar Kraus. Two volumes. Leipzig: Felix Meiner Verlag, 1924/1925.
Dantzig, Tobias. *Number: The Language of Science*. New York: MacMilaan, 1939.
Eves, Howard and Newson, Carroll V. *An Introduction to the Foundations and Fundamental Concepts of Mathematics*. Revised edition. New York: Holt, Rinehart and Winston, 1965.
Frege. Gottlob. *The Foundations of Arithmetic. A logico-mathematical enquiry into the concept of number*. Translated by J.L. Austin. Oxford: Basil Blackwell, 1968.
Klein, Jacob. 'The Concepts of Numbers in Greek Mathematics and Philosophy.' Paper delivered before the Philosophy Club of the University of Virginia, 6 March 1939. Mimeographed.
– *Greek Mathematical Thought and the Origin of Algebra*. Translated by Eva Brann. Cambridge, Mass.: The M.I.T. Press, 1968.
Kline, Morris. *Mathematics in Western Culture*. New York: Oxford University Press, 1953.
Körner, Stephan. *The Philosophy of Mathematics*. London. Hutchinson and Co., 1971.
Martin, Gottfried. *Klassische Ontologie der Zahl. Kantstudien*, Ergänzungsheft 70 (1956).
Nagel, Ernest and Newman, James R. *Goedel's Proof*. New York: New York University Press, 1968.
Russell, Bertrand. *Introduction to Mathematical Philosophy*. London: Allen and Unwin, 1976.
Waismann, Friedrich. *Introduction to Mathematical Thinking*. Translated by T.J. Benac. New York: Frederick Ungar, 1951.

INDEX

Abacus, 82
Absence, see Presence and absence
Abstraction, 35, 37, 42, 91ff., 106, 132
  ideational, 94f.
  idealizing, 105
Aggregate, 3, 33, 38, 42, 54, 55, 57, 60, 68, 70, 85
Algebra, 23, 110ff., 114
Algorithm, 14, 15, 16f., 18, 20, 125
Analysis, 1 ff., 89, 104, 109, 120ff., 130, 135, 137
  arithmetization of, 2 ff., 15
  clarification of, 15, 18, 120ff., 135
  justification of, 15, 125ff., 135ff.
'And,' meaning of, 72, 86
Angelelli, I., 27
Apophantics, 101, 107, 113f., 120f., 136f.
Aristotle, 86, 104, 106, 131
Arithmetic, 1, 6, 9, 12, 13, 14, 16, 18, 89ff., 100ff., 107, 108, 109ff., 122, 125, 135, 136
  universal, 3, 11, 20, 84, 110f., 130
Austin, J.L., 41

Bachelard, S., 27, 131, 134
Becker, O., 23, 28, 44
Being, 35f., 42, 77, 86, 87, 90, 105, 118, 120, 132, 137
Benac, T.J., 132
Benacerraf, P., 105
Berkeley, G., 2
Bernays, P., 105
Beth, E.W., 104
Biemel, W., 25, 26, 27, 43
Bosio, F., 28
Boyce-Gibson, W.R., 26, 27, 86, 131
Boyer, C.B., 23, 24, 130
Brann, E., 42
Brentano, F., 3, 4, 7, 9, 13, 15, 19, 20, 25, 26, 28, 43f.

Cairns, D., 27, 131
Calculation, 80ff., 87, 102f., 108, 111, 112, 115f., 118, 131, 136
Calculus, 1, 2, 20, 89, 109, 123
Cantor, M., 24, 131
Carr, D., 27, 131
Categorial intuition, 73, 86, 94
Categorial objects, 36, 37, 43, 45, 61, 99, 136
Collecting, 18, 36, 37, 46, 54, 71f., 75, 91, 93f., 97, 115, 118, 121
Collective connection, 70ff., 86, 91, 94
Comparison, 55ff., 62, 65, 68, 76, 81, 136
Completeness, 128, 134, 135
Constitution, 35f., 37f., 42, 43, 46, 90
Counting, 3, 4, 8, 9, 16, 18, 21, 36, 37, 43, 46, 98, 118, 121, 136
  authentic, 37, 41, 43, 45, 50, 59, 60, 65, 66ff., 71f., 75, 76f., 78, 79, 84, 93f.
  'everyday,' 37, 43, 58f., 80, 82ff., 87, 136

D'Alembert, J., 2
Dantzig, T., 106
De Boer, T., 25, 26, 38, 42, 43
Dedekind, R., 6, 24, 121
Definiteness, 128, 135
Descartes, R., 130
Diophantus, 132, 134

Eley, L., 23, 24, 25, 134
Elveton, R.O., 23, 25
Embree, L.E., 131
Entwistle, W.J., 60
Equality and inequality, 31, 40, 55ff., 62, 65, 92, 94, 102, 136
Eudoxus, 131
Euclid, 17, 31ff., 41, 42, 116, 122, 133
Eves, H., 23
Evidence, 35, 118, 121, 122, 123, 137
Explication, 51ff., 55, 56, 58, 59, 61, 62, 65, 68, 76, 81, 136

Facts, 36, 46
Farber, M., 25, 27
Figural moments, 48ff., 50, 52, 60
Findlay, J.N., 26, 29, 131
Føllesdal, D., 27
Frege, G., 19, 21, 27, 31ff., 41, 42, 58, 85, 104, 105

Gauss, C.F., 25, 27
Geometry, 2, 5, 8, 23, 29, 100, 103, 111, 116f., 118, 122, 133
Ghetaldi, M., 132
Goethe, J.W., 19
Goedel, K., 128, 134
Gurwitsch, A., 61

Hankel, H., 118, 132
Harriot, T., 132
Helmholtz, H. v., 84
Heath, T.L., 41
Hilbert, D., 128, 134
Huertes-Jourda, J., 28
Husserl, E., works by:
  Cartesianische Meditationen, 46, 49
  'Doppelvortrag,' 16, 18, 121, 126
  Erfahrung und Urteil, 46, 53, 60, 61, 74, 75, 76, 96, 99
  Formale und transzendentale Logik, 16, 18, 22, 26f., 37, 74, 76, 106, 118, 119, 120f., 125, 128, 129, 133, 135
  Ideen zu einer reinen Phänomenologie und phänomenologischen Philosophie, 16, 26, 27, 74, 76, 110, 129
  Krisis der europäischen Wissenschaften und die transzendentale Phänomenologie, 16, 27, 110, 129
  Logische Untersuchungen, 16, 19ff., 23, 26, 28, 29, 48, 50, 72ff., 75, 76, 77f., 86, 87, 89, 90, 92ff., 104, 105, 106, 108, 111, 114, 120, 126, 135
  Philosophie der Arithmetik, 1, 9, 10ff., 15, 17f., 19ff., 24, 27, 37ff., 46ff., 50, 55ff., 61, 70, 72, 75, 76f., 80ff., 85, 86, 87, 90ff., 99, 101ff., 105, 106, 107, 108, 110, 114, 125f., 133, 135
  'Über den Begriff der Zahl,' 4ff., 11, 20ff., 39f., 42, 44, 55, 86, 105
  'Ursprung der Geometrie,' 34, 79f., 82, 87

Ideograms, 87
Imaginary, problem of, 18, 126ff., 134
Intentions, empty and filled, 35, 38, 39, 42, 43, 44, 45, 67, 72, 78, 86, 89, 93, 95f., 105, 135; see also Presence and absence; Presentation, authentic and symbolic
Inverse operations, 111

Janssen, P., 26

Kindinger, R., 26
Klein, J., 42, 85f., 104, 107, 131, 132, 133
Kline, M., 23
Kronecker, L., 23
Landgrebe, L., 28, 29, 44
Leibniz, G.W., 5, 130
Locke, J., 106
Logic, 13f., 15, 17, 19, 20, 25, 26, 29, 101
    formal, 13f., 17, 26, 113
    truth-logic, 119f.

Mach, E., 60
Manifolds, 16, 27, 115ff., 124
Manifold theory, 4, 16, 18, 109, 114, 115ff., 120, 121ff., 126f., 129, 132, 133
Martin, G., 104, 133
Meinong, A., 26
Metaphysics, 25
Mill, J.S., 28, 61, 106
Multitudes, 32ff., 39, 40, 42, 44, 54, 55, 56, 57, 58, 60, 61, 62, 67, 68, 70, 83, 85, 86, 90f.

Nagel, E., 134
Naming, 65f., 79ff., 116, 118, 122, 137
Natorp, P., 24
Newman, J.R., 134
Newsom, C.V. 23
Newton, I., 1, 2, 5, 130
Nominalism, 84
Nominalization, 74ff., 78
Number:
  ancient Greek view of, 42, 85f., 107, 130, 133
  complex, 12, 126
  concept of, 3, 4, 6, 7, 8, 9f., 11, 18, 32f., 62, 131
  definition of, 31ff., 41
  formal character of, 98ff., 110, 130
  fractional, 15, 27, 124, 125
  German terms for, 42
  ideal, 89f., 93ff., 110, 121, 136
  imaginary, 5, 10, 15, 27, 125, 134
  irrational, 2, 10, 12, 13, 15, 27, 122, 124, 125ff., 134
  large whole numbers, 9f., 13, 15, 59, 66, 76ff., 87, 102, 136
  negative, 2, 10, 12, 13, 15, 27, 122, 124, 125ff., 134
  objectivity of, 7f., 21f., 29, 45, 71
  'one' as, 27, 33, 67, 69, 85, 133, 136
  ordinal, 12f.
  origin of, 4, 8, 11, 18, 19, 20, 28, 34, 36,

37, 39ff., 43, 45, 55, 59f., 90ff., 99, 106
positive whole, 2, 6, 122, 123, 126, 127f., 135
rational, 2, 10, 12, 13, 122, 124, 125ff.
real, 2, 16, 42, 109, 120, 126, 133, 135
Scholastic view of, 106
unity of, 69ff., 76ff., 85f.
'zero' as, 27, 33, 67, 69, 85, 123, 133, 136

One-to-one correspondence, 57, 83
Ontology, 42, 107, 109, 118, 123, 126f., 129, 133
formal, 101ff., 107, 109, 113, 118, 136
regional, 100
Origin, 34ff., 43; see also Number, origin of
Oughtred, W., 132

Pairing, 46, 49
Passive givenness, 45ff., 53, 60, 66
Picker, B., 24, 25
Pietersma, H., 29, 41, 86
Pivcevic, E., 27, 41
Plato, 85f., 87, 90, 104f., 107, 130
Presence and absence, 35f., 39, 41, 42, 45, 50, 55, 59, 65ff., 69, 72, 75, 76, 77ff., 80, 82, 83f., 117, 121, 122, 123, 124, 125, 135f., 137
Presentation, 19, 22, 25, 28, 34, 99
authentic and symbolic, 9, 12, 14, 18, 21f., 28, 36, 38f., 40f., 42, 43, 57, 61, 71, 76, 78, 87, 89, 95, 102, 118; see also Presence and absence; Intentions, empty and filled
Psychologism, 8, 19ff., 27, 28, 29
Psychology, 7, 8, 11, 14, 19, 21, 25, 27, 28, 29, 34, 42, 61, 137
descriptive, 7, 8, 19, 21, 25, 28
Putnam, H., 105
Pythagoreans, 85, 132

Quasi-numbers, 15, 18, 125f.
Quasi-qualitative moments, 48ff., 60, 61, 62

Retention, 53ff., 56ff., 62, 67, 77f.
Riemann, G.F., 27, 116f., 131
Riese, A., 17
Russell, B., 106

Schuhmann, K., 23, 24, 26, 27, 28, 43
Sedimentation, 79ff., 87, 102, 103, 119, 125, 136f.
Sense data, 48ff., 61, 73
Sensuous groups, 45ff., 51, 52, 54, 55, 59, 60, 61, 62, 65, 66, 68, 69, 77
Sigwart, C., 28
Sokolowski, R., 28, 42f., 62, 85, 87, 104f., 106, 107, 132, 133
'Something,' concept of, 98ff., 106
Spiegelberg, H., 23, 25
Stolz, O. 58
Stumpf, C., 4, 12, 13, 17

Theory forms, 112f., 115, 122, 136
Torretti, R., 28, 43, 62, 86, 87
Truth, 35, 113, 117, 119f.
Tugendhat, E., 36, 43, 44, 86, 106

Units, 32, 55

Verdenal, R., 26
Vieta, F., 110, 130, 132, 134
Voltaire, F., 2
Von Ehrenfels, C., 60

Waismann, F., 87, 132
Weierstrass, K., 2ff., 11f., 15, 17, 19, 20, 23, 24, 31, 41
Willard, D., 24, 25, 26, 131
Winance, E., 107

GPSR Compliance

The European Union's (EU) General Product Safety Regulation (GPSR) is a set of rules that requires consumer products to be safe and our obligations to ensure this.

If you have any concerns about our products, you can contact us on ProductSafety@springernature.com

In case Publisher is established outside the EU, the EU authorized representative is:

Springer Nature Customer Service Center GmbH
Europaplatz 3
69115 Heidelberg, Germany

Batch number: 10823959

Printed by Printforce, the Netherlands